THE LONG JOURNEY

THE LONG JOURNEY
Exploring Travel and Travel Writing

Edited by
Maria Pia Di Bella
Brian Yothers

berghahn
NEW YORK • OXFORD
www.berghahnbooks.com

Published in 2021 by
Berghahn Books
www.berghahnbooks.com

© 2021 Berghahn Books

Originally published in *Journeys:
The International Journal of Travel and Travel Writing*

All rights reserved. Except for the quotation of short passages for the purposes of criticism and review, no part of this book may be reproduced in any form or by any means, electronic or mechanical, including photocopying, recording, or any information storage and retrieval system now known or to be invented, without written permission of the publisher.

Library of Congress Cataloging-in-Publication Data

Names: Di Bella, Maria Pia, editor. | Yothers, Brian, 1975- editor.
Title: The long journey : exploring travel and travel writing / edited by Maria Pia Di Bella, Brian Yothers.
Description: First Edition. | New York : Berghahn Books, 2020. | "Originally published in Journeys: The International Journal of Travel and Travel Writing"--T.p. verso. | Includes bibliographical references and index.
Identifiers: LCCN 2020031454 (print) | LCCN 2020031455 (ebook) | ISBN 9781789209358 (Hardback) | ISBN 9781789209365 (Paperback) | ISBN 9781789209372 (eBook)
Subjects: LCSH: Travel writing. | Voyages and travels. | Travel writers.
Classification: LCC G151 .L664 2020 (print) | LCC G151 (ebook) | DDC 910.4--dc23
LC record available at https://lccn.loc.gov/2020031454
LC ebook record available at https://lccn.loc.gov/2020031455

British Library Cataloguing in Publication Data
A catalogue record for this book is available from the British Library

ISBN 1-78920-935-8 hardback
ISBN 1-78920-936-5 paperback
ISBN 978-1-78920-937-2 ebook

Contents

List of illustrations — vii

Introduction — 1
Maria Pia Di Bella and Brian Yothers

Part I. Memory and Trauma

Chapter 1
Walking Memory: Berlin's 'Holocaust Trail' — 13
Maria Pia Di Bella

Chapter 2
Touring the African Diaspora — 28
Cheryl Finley

Chapter 3
A Wartime Cinematic Recreation of the Journey Linking China and Japan in the Modern Era — 35
Joshua A. Fogel

Part II. Visualizing Otherness

Chapter 4
Seeing a Difference: Spectacles of Otherness in Eighteenth-Century Illustrated Travel Books — 53
Julia Thomas

Chapter 5
A Beginning, Two Ends, and a Thickened Middle: Journeys in Afghanistan from Byron to Hosseini — 70
Graham Huggan

Contents

Chapter 6
New Men, Old Europe: Being a Man in Balkan
Travel Writing — 85
Wendy Bracewell

Chapter 7
Among Cannibals and Headhunters: Jack London
in Melanesia — 108
Keith Newlin

Part III. Creating and Recovering Perspective

Chapter 8
Forgetting London: Paris, Cultural Cartography,
and Late Victorian Decadence — 133
Alex Murray

Chapter 9
In The Eyes of Some Britons: Aleppo, an
Enlightenment City — 150
Mohammad Sakhnini

Chapter 10
An Ordinary Place: Aboriginality and
'Ordinary' Australia in Travel Writing of the 1990s — 168
Robert Clarke

Chapter 11
The Right Sort of Woman: British Women
Travel Writers and Sports — 188
Precious McKenzie Stearns

Conclusion — 201
Pramod K. Nayar

Index — 206

List of Illustrations

1.1:	Tourists around the Empty Library in Bebelplatz (Berlin, 2011).	15
1.2:	Gunter Demnig discussing with the people who ordered the Stolpersteine where to put the Stolpersteine.	20
1.3:	Gunter Demnig and his aide working in order to insert the six Stolpersteine in the macadam.	21
1.4:	The six Stolpersteine inserted in the macadam surrounded by flowers.	21
1.5:	Steglitz, the Spiegelwand.	22
1.6:	18-10-1941: First train from Berlin to Lodz, carrying 1251 Jews.	23
1.7:	Berlin-Grunewalld, "Gleis 17." Last train going from Berlin to Theresienstadt with 18 Jews on board.	23
2.1:	*Description of a Slave Ship*, London Committee of the Society for the Abolition of the Slave Trade, engraving by James A. Phillips, 1789.	32
4.1:	Bernard Picart, frontispiece for Cornelius Le Bruyn, *Travels into Muscovy, Persia, and Part of the East Indies*, vol. 1, London, 1737.	54
4.2:	William Hodges, *Family in the Dusky Boy, New Zealand*, from James Cook, *A Voyage towards the South Pole and round the World*, vol. 1, London, 1777.	59
4.3:	*The Landing of Captain Roger's men at California, and their reception by the Natives*, Edward Cavendish Drake, *A new universal collection of authentic ... Voyages and Travels*, London, 1768.	62
7.1:	"Man-Eaters." From *The Cruise of the Snark* (1911), 260.	117

7.2: "Cannibal Bushmen at Forte, Northwest Malaita." From "Cruising in the Solomons," *Pacific Monthly* 23 (June 1910), 593. 117

7.3: "A Native South Sea Island Missionary, in What He Considers Correct Clerical Garb." From "Stone Fishing at Bora-Bora," *Pacific Monthly* 23 (April 1910), 342. 123

7.4: "Laundry Bills Are Not Among His Vexations. His Garb, However, Is a Concession to Civilization.—Lord Howe Atoll. From *The Cruise of the Snark*, (1911), 332. 124

7.5: "A Prince of Polynesia," JLP 502, Alb. 64, Jack London Papers. 124

7.6: "The Two Handsomest Men in All the Solomons," From "Cruising in the Solomons," *Pacific Monthly* 24 (July 1910), 38. 125

7.7: "One Oreinte (Inca) Andes Indian reduced dried human." 125

INTRODUCTION

Maria Pia Di Bella and Brian Yothers

The study of travel is inherently interdisciplinary, bringing together a host of practices and narratives embedded in history, culture, and society. The chapters in this volume illustrate the range and vibrancy of scholarly work associated with travel: the intersection of travel with historical and communal memory appears in essays on the Holocaust in Europe and memories of slavery in Africa and war in Asia; the representation of otherness informs essays on topics ranging from illustrations in British colonial travel writing to travel in Afghanistan, the Balkans, and Melanesia; and the varied perspectives from which travel can be viewed appears in essays on Paris, Aleppo, Argentina, and Australia. This collection thus circles the globe even as it takes in multiple disciplinary perspectives, often within the same chapter. The taxonomy that we as the editors have used to divide the collection reflects how travel and travel writing can be read via historical memory and trauma (part I), the representation of otherness (part II), and the reimagining of perspective (part III). Notably, each of these essays engages with questions raised by all of the three sections, illustrating the interpenetration of major themes within travel writing.

The study of travel and travel writing has developed across disciplinary and geographical boundaries, at once taking in nearly the entire planet as well as the full range of disciplines across the humanities and social sciences. In social sciences like anthropology and sociology, the emphasis has been on how meaning is created through the act of travel itself, even as critical analysis of writing about travel has served as a necessary component to discussions of the practice of travel, beginning with Claude Lévi-Strauss's *Tristes Tropiques* (1955). Victor and Edith Turner's work on pilgrimage in *Image and Pilgrimage in Christian Culture* (1978), for example, has helped to set the tone for a great deal of anthropological work on travel with its emphasis on pilgrimage as a liminal space, and the influence of their model has also worked its way into literary and historical studies related to travel. Maria Pia Di Bella, co-editor of this volume and co-founder of *Journeys*[1] has captured the range of significance that can be associated with travel as practice and lived experience in her work, included in this volume, on the Holocaust trail in Berlin in the early twenty-first century.

Notes for this section can be found on page 8.

Maria Pia Di Bella and Brian Yothers

In literature and history, travel has often been most closely connected to the study of empire. Edward Said's *Orientalism* (1977) is a foundational text for this sort of work on travel writing, as Said sought to show how European travelers created an imagined Orient that they were able to control, in part as a result of the narratives they crafted. Mary Louise Pratt offered a similar focus on how travel, travel writing, and empire could intertwine with her 1992 study *Imperial Eyes: Travel Writing and Transculturation*, which offered a reading of travel writing that focused on how the "gaze," raced as white and gendered as male, could become a mode of power and domination. If Said had emphasized European imperialism in Africa and Asia (especially British and French imperialism), Pratt emphasized the role of travel and travel writing in New World colonialism and imperialism. The interconnectedness of travel and empire also appeared in the New Historicist scholarship of Stephen Greenblatt, notably in his essay "Invisible Bullets" (1988), which showed how intellectual history, literature, and mobility could come together in one extraordinary essay. More recently, Pramod Nayar, who has provided the afterword for this volume, has offered substantial discussions of British travel writing in the Indian subcontinent in *English Travel Writing in India, 1600-1920: Colonizing Aesthetics* (2007) and of Indian travelers in Europe in *Indian Travel Writing in the Age of Empire, 1830-1940* (2020), and the work of William Dalrymple has drawn on travel and travel writing as Dalrymple has narrated the history of British India across multiple studies.

As the study of travel writing has developed as a field, scholars have become interested in capturing more equivocal dimensions of travel than those emphasized by Said and Pratt. Paul Fussell's *Abroad* (1980), for example, considered British travel writing as a source of the cultural and literary vibrancy of the 1920s and 1930s, and in James Buzard's 1993 study *The Beaten Track: European Tourism, Literature, and the Ways to Culture, 1800-1920*, Buzard was less interested in travel as an expression imperial ambition than in the ways that Europeans constituted distinctive modes of travel: tourism versus more serious, cerebral modes of travel, for example, and authenticity of experience versus illusion. Graham Huggan and Peter Holland, meanwhile, have explored more recent approaches to travel, again with an emphasis on the generative aspects of travel writing, in their 1999 study *Tourists with Typewriters: Critical Reflections on Contemporary Travel Writing*. Peter Hulme and Tim Youngs's 2002 *Cambridge Companion to Travel Writing* served to illustrate the geographical range of travel writing while also inviting discussions of travel writing in relation to the major preoccupations of cultural studies at the turn of the twenty-first century. Hulme and Youngs thus contributed to bridging the gap between considerations of travel writing as colonial representation and considerations of travel writing as a representation of the self.

Introduction

Brian Yothers, a co-editor of this volume and of *Journeys*, has written about travel in contexts that combine the strand of travel writing research concerned with authenticity with the strand concerned with questions of empire in his work on nineteenth-century US travelers to Ottoman Palestine in his study *The Romance of the Holy Land in American Travel Writing, 1790-1876* (2007). This particular strand of travel, with its mix of cultural, political, and aesthetic questions, has inspired multiple studies, with Hilton Obenzinger's *American Palestine* (1999), Bruce Harvey's *American Geographies* (2001), John Davis's *The Landscape of Belief* (1996), Burke O. Long's *Imagining the Holy Land* (2002), and Jeffrey Alan Melton's *Mark Twain, Travel Books, and Tourism* (2002) investigating US travelers in the Middle East, Asia, and the Americas. Gender has joined race as a crucial mode of analysis in travel writing studies, as in Malini Johar Schueller's *US Orientalisms* (1998), which brought together questions of colonial representation and gender in relation to US travel writing. The essays in the volume thus emerge against the backdrop of a vigorous and growing field that has incorporated elements of literary and historical studies, anthropology and sociology, and the analysis of practice or representation across a wide range of media.

Memory and Trauma

The opening portion of *The Long Journey* illustrates how travel as a practice and travel writing as a mode of representation interact with historical and communal memory. Each essay both corresponds to the experience of a distinct region (Europe, Africa, East Asia) and to the complexities of trauma, guilt, and memory. Each shows how the act of travel itself can become a significant part of the practice of memory. The essays in this section also work across multiple media: the design of museums and memorials, the construction of prose narratives, and the role of cinema in representations of travel.

In "Walking Memory: Berlin's 'Holocaust Trail,'" Maria Pia Di Bella explores how the city of Berlin has developed a Holocaust trail of street memorials that become sites for secular pilgrimage and an communal remembrance. As she points out, the choice to identify multiple places of remembrance for the Shoah around the city constitutes a choice that German governments made after reunification, as the number of sites identified over the decades before reunification were much sparser. The essay thus illustrates how travel as practice is shaped by forms of representation even as it shapes them, and how these representational choices are shaped by historical developments, most notably in this case the transformation of German national self-understanding after reunification. Poems, sculptures, and art installations are contribute to

3

historical memory, Di Bella shows, along with specific sites of violence associated with the Holocaust and the train stations used to carry out genocide.

In "Touring the African Diaspora," Cheryl Finley considers heritage tourism in Africa and the relationship between this form of travel and the task of remembering the historical trauma associated with slavery. Finley shows how recent work in the visual arts has worked to create a memorial that has otherwise been denied to the descendants of enslaved people. Taking Toni Morrison's *Beloved* as a point of departure, Finley plays literary texts, works of visual art, and practices of tourism off on one another as she presents a taxonomy of the relationship between Africa and the Americas. For example, Finley examines art installations in the Americas and England and compares them to the original sites in West Africa on which the installations are based. The circulation of memories of place and visual images across the Atlantic and across the centuries thus becomes central for Finley's essay, again illustrating the interpenetration of travel as practice and representation.

In "A Wartime Cinematic: Recreation of the Journey Linking China and Japan in the Modern Era," the historian Joshua A. Fogel begins with the voyage of a ship called the *Senzaimaru* "of investigation and trade" from Japan to China in the 1862 and proceeds to consider how that voyage has been represented in film in the twentieth century. Fogel points to a 1944 cinematic representation of the voyage of the *Senzaimaru* that was found in the former Soviet film archives in 2001. As Fogel establishes, this film was created in Japanese-occupied Manchuria during World War II, and it found its way into Soviet archives after Japan's defeat in the war. Unexpectedly, Fogel shows that even though this film was created in the context of Japanese war-time propaganda, it actually reflected a genuine sense of cultural exchange against all odds. Fogel's essay leads into the preoccupations that shape the essays in the second portion of this collection, which are concerned with representations of otherness, particularly in the context of complex power relations. Fogel offers a fascinating account of how history and the arts continue to intersect over time through the temporal and physical voyages of people and artifacts.

Visualizing Otherness

The second cluster of essays, "Visualizing Otherness" addresses matters of vision and representation, taking up the concern over otherness that so shaped the seminal work of Edward Said and Mary Louise Pratt in providing a theoretical basis for travel writing studies.

Julia Thomas's essay, "Seeing a Difference: Spectacles of Otherness in Eighteenth-Century Illustrated Travel Books," investigates a period at which modern travel writing was developing as a genre. As with walking tours of cities and cinematic and novelist representations, there has always been a multi-media dimension to how travel has been represented. Thomas takes us on a tour through a wide range of eighteenth-century travel books, showing how the images in each of them are characterized by both "visual delights and troubling inconsistencies." The images that Thomas explores offer a tour of the eighteenth-century world as experienced by travelers, but for twenty-first century readers, it also offers a fascinating tour of the mentality of eighteenth-century Europeans and how their mental picture of the world became manifest in illustrated travelogues. As with Di Bella and Finley, Thomas emphasizes the centrality of the visual in the representation of travel across media, and here the visionary, imaginative aspects of travel writing are in tension with travel's contribution to historical memory.

In "A Beginning, Two Ends, and a Thickened Middle Journeys in Afghanistan from Byron to Hosseini," Graham Huggan takes on the question of how travel is visualized in conditions of peril. As with the first three essays in the volume, Huggan works across centuries, considering how wars in Afghanistan were represented in the travel literature of the twentieth century and the early twenty-first century. Huggan considers the mis-representation of Afghanistan as part of the "Greater Middle East" as part of the propaganda in support of the War on Terror of the early 2000s, and he shows how looking at travel narratives diachronically can provide a useful corrective to misrepresentations. Here, as in the first three essays in the volume, questions of historical memory are paramount; for Thomas, the point that travel writing is a product of imagination and preconception as well as observation is crucial. For Huggan, old (twentieth-century British) and new (twenty-first-century American) forms of imperialism intersect in representations of the European and North American travelers' gaze on Afghanistan over time.

In "New Men, Old Europe: Being a Man in Balkan Travel Writing," Wendy Bracewell investigates a similarly complex context, as she considers how gender, and specifically masculinity, connects with the matter of travel in zones of conflict and threat. Bracewell points out that masculine toughness has been central to travelers' representations of southeastern Europe, and she uses Said's discussion of a feminized Orient as a foil for her consideration of a masculinized Balkan region. Bracewell considers Maria Todorova's (1997) contention that this emphasis of masculinity in "Balkanism" reflects the "imputed ambiguity" of the Balkans' status within Europe, but concludes that the various alternative "solidarities" that male and female travelers might embrace make it difficult to create a straightforward taxonomy of power relations based on the gender of travelers or of those they represent. Bracewell takes as her subject matter the emergence of a rich array of travel accounts of southeastern

Europe in the years after the fall of the Soviet bloc. She demonstrates that the representations of masculinity that tie these accounts together are profoundly historically conditioned, both in terms of the culture of the traveler and the culture of the destination.

Part II is rounded off by a richly visual piece that brings together questions of identity, otherness, and perception that have shaped all the essays in this section. In "Among Cannibals and Headhunters: Jack London in Melanesia," Keith Newlin examines the writings of a figure who has often been read as a dissident from his culture's racism and imperialism. Reading photographic images associated with London's voyage and the artifacts London collected along with London's travel writing, Newlin shows that these readings of London have missed the substantial degree to which London accepted and internalized racism in his responses to Melanesian islanders. This chapter illustrates how the early discussions of empire in travel writing studies can be complicated by the individual author whose work is explored, but also how authors who are conventionally regarded as being exceptional may be more representative of their own culture's myopia than is commonly recognized.

Creating and Recovering Perspective

In part III of this volume, "Creating and Recovering Perspective," the final cluster of essays shows the significance of angles of vision to travel and travel writing, moving from accounts of travel within Europe to western Asia, Europe, and finally, Argentina and India.

In "Forgetting London: Paris, Cultural Cartography, and Late Victorian Decadence," Alex Murray examines the representations of London that emerged in late Victorian writing. Murray shows how Paris came to shapes writers' perspectives on London through the artistic conventions of Naturalism and Impressionism, thus bringing a painterly perspective to bear on the representation of the cityscape. Using the British Decadent novelist George Moore's dictum that "To write about London I should have to begin by forgetting Paris," Murray shows that the visual arts provide a grammar for how late nineteenth-century British writers imagined their own capital city. Murray shows that the change in perspective provided by the time that Moore and his fellow Decadent writer Arthur Symons spent in Paris meant that they were able to change how London was represented in the decades that followed. Here, then, travel as practice brings writers into contact with new forms of representation, which in turn reshape the gaze that they direct toward places formerly familiar to them. For Murray, travel becomes an enabling condition for new visions of place, including the home to which the traveler returns.

Introduction

That Orientalism as a concept continues to shape studies of travel writing is evident in Mohammad Sakhnini's essay, "In The Eyes of Some Britons: Aleppo, an Enlightenment City," which considers early British narratives about Syria's second city. Sakhnini explicitly pairs the eighteenth-century British response to Aleppo with contemporary anti-immigrant sentiment in the United Kingdom, showing how eighteenth-century British travelers recognized the complexity of the Syrian cultural and intellectual milieu, and indeed presented Aleppo as a model of the Enlightenment values that the travelers hoped to realize in their own cities. Sakhnini's essay thus complicates the discussions of Orientalist travel that have been in circulation since Said's *Orientalism*, and Sahhnini reflects movingly on the way in which Aleppo's historical trajectory—and London's—reflects the fragility of all our social and cultural arrangements and the precariousness of human experience.

Robert Clarke's essay, "An Ordinary Place: Aboriginality and 'Ordinary' Australia in the Travel Writing of the 1990s," points to the ways in which travel writers represent and mis-represent indigenous Australians. Clarke surveys as range of late twentieth-century texts, highlighting how authors who engaged with aboriginality in Australia arrived at perspectives that avoided the conventional exoticism and condescension of much the sort of travel literature that is built around the imperial gaze that Mary Louise Pratt identified. In particular, Clarke shows that by regarding Aboriginal Australia as "ordinary," as not defined by otherness, at least some travel writers are establishing an alternative to modes of travel writing that have been shaped and warped by colonialism.

In "The Right Sort of Woman: British Women Travel Writers and Sports," Precious McKenzie Stearns examines travel from a vantage point that has attracted less attention than it deserves: that of travel's connection to the presence of human bodies, and specifically women's bodies, in the natural world. Stearns shows that women who engaged in big-game hunting in Argentina and India in the nineteenth century offered alternative perspectives on femininity to those that conventionally shaped travel literature in the era of British colonialism. In this essay, humans become participants in as well as observers of the landscape, with profound implications for how we understand the ecology of travel as well as the role of gender in shaping travelers' self-representations.

These chapters constitute a cross-section of the finest essays to appear over the last 20 years in *Journeys*. Some difficulties along the way in the production process meant that we were not able to include every essay that we wanted to in the print volume, and so these essays have been made available in their original form for free on the Berghahn Books website. These essays are Amardeep Singh's "Veiled Strangers: Rabindranath Tagore's America, in Letters and Lectures" (2009), Nigel Rapport's "Walking Auschwitz, Walking without Arriving" (2008), Andrew Irving's "Journey to the End of Night: Disillusion and Derangement among the Senses" (2008), and Brian Yothers's "Facing East, Facing West: Mark Twain's *Following the Equator* and Pandita Ramabai's *The Peoples of the United States*" (2009). All four of these essays are available in the archives of *Journeys* for free as a supplement to the essays printed in this volume. The editors would like to invite the readers of this volume to consider these pieces as an extension of the work that is contained in this printed volume.

If there is one lesson that the varied and provocative essays included in this volume and on the Berghahn site can teach us, it is the remarkably protean quality of travel and the cultural productions associated with its practice. We travel for reasons that cover the range of human experience: pleasure and need, exploration and self-discovery, in order to write and in order to remember. The essays in this volume illustrate how travel and travel writing function across a range of motivations, practices, and representational devises, and they emphasize the vastness of the topic of journeys and the attempt to describe and narrate them.

Maria Pia Di Bella is co-founder and co-editor of the journal *Journeys: The International Journal of Travel and Travel Writing*. She is a senior research scholar at IRIS-EHESS, Paris, and Research Affiliate at the Harvard Divinity School. She has published monographs on popular religions and cultures of punishment and penitence, including *La Pura verità. Discarichi di coscienza intesi dai Bianchi* (1999), *Dire ou taire en Sicile* (2008), and *Essai sur les supplices. L'état de victime* (2011). She has edited, among others, *Vols et sanctions en Méditerranée* (1998) and with James Elkins, *Representations of Pain in Art and Visual Culture* (2012). Since 2011 she works on street memorials in Berlin concerning the victims of genocide and the function of these memorials as "symbolic reparations."

Brian Yothers is the Frances Spatz Leighton Endowed Distinguished Professor of English and the Chair of the Department of English at the University of Texas at El Paso. He is co-editor of the journal *Journeys: The International Journal of Travel and Travel Writing* and the author or editor of 13 books and special issues of journals.

Notes

1. *Journeys: The International Journal of Travel and Travel Writing*, came out in 2000, published by Berghahn Books.

    Editorial board:
    Founding Editor: Garry Marvin (University of Surrey Roehampton, London)
    Robert C. Davis (University of Ohio)
    Maria Pia Di Bella (CNRS, Paris)
    John Eade (University of Surrey Roehampton, London)

References

Buzard, James. 1993. *The Beaten Track: European Tourism, Literature, and the Ways to Culture, 1800-1920*. New York: Oxford University Press.
Davis, John. 1996. *The Landscape of Belief*. Princeton, NJ: Princeton University Press.
Fussell, Paul. 1980. *Abroad: British Literary Traveling Between the Wars*. New York: Oxford University Press.
Greenblatt, Stephen. 1988. "Invisible Bullets." In *Shakespearean Negotiations*, 21-39. Berkeley, CA: University of California Press.
Harvey, Bruce A. 2002. *American Geographies: U.S. National Narratives and the Representation of the Non-European World, 1830-1865*. Stanford, CA: Stanford University Press.
Huggan, Graham, and Peter Holland. 1999. *Tourists with Typewriters: Critical Reflections on Contemporary Travel Writing*. Ann Arbor: University of Michigan Press.
Hulme, Peter, and Tim Youngs. 2002. *Cambridge Companion to Travel Writing*. New York: Cambridge University Press.
Irving, Andrew. 2008. "Journey to the End of Night: Disillusion and Derangement among the Senses." *Journeys: The International Journal of Travel and Travel Writing*. Vol. 9, no. 2, 138-160.
Levi-Strauss, Claude. 1955, English trans. 1961. *Tristes Tropiques*. Trans. John Russell. London: Hutchinson & Co.
Long, Burke O. 2002. *Imagining the Holy Land: Maps, Models, and Fantasy Travels*. Bloomington, IN: Indiana University Press.
Melton, Jeffrey Alan. 2002. *Mark Twain, Travel Books, and Tourism*. Tuscaloosa, AL: University of Alabama Press.
Nayar, Pramod. 2007. *English Travel Writing in India, 1600-1920: Colonizing Aesthetics*. New York, Routledge.
Nayar Pramod. 2020. *Indian Travel Writing in the Age of Empire, 1830-1940*. New York: Bloomsbury.
Obenzinger, Hilton. 1999. *American Palestine: Melville, Twain, and the Holy Land Mania*. Princeton, NJ: Princeton University Press.
Pratt, Mary Louise. 1992. *Imperial Eyes: Travel Writing and Transculturation*. New York: Routledge.
Rapport, Nigel. 2008. "Walking Auschwitz, Walking without Arriving." *Journeys: The International Journal of Travel and Travel Writing*. Vol. 9, no. 2, 33-54.
Said, Edward. 1977. *Orientalism*. New York: Columbia University Press.
Schueller, Malini Johar. 1998. *U.S. Orientalisms: Race, Nation, and Gender in Literature, 1790-1890*. Ann Arbor, MI: University of Michigan Press.
Singh, Amardeep. 2009. "Veiled Strangers: Rabindranath Tagore's America, in Letters and Lectures." *Journeys: The International Journal of Travel and Travel Writing*. Vol. 10, no. 1, 51-68.
Turner, Victor and Edith. 1978. *Image and Pilgrimage in Christian Culture*. New York: Columbia University Press.
Yothers, Brian. 2009. "Facing East, Facing West: Mark Twain's *Following the Equator* and Pandita Ramabai's *The Peoples of the United States*." *Journeys: The International Journal of Travel and Travel Writing*. Vol. 10, no. 1, 107-119.
Yothers, Brian. 2007. *The Romance of the Holy Land in American Travel Writing, 1790-1876*. Aldershot, UK: Ashgate.

Part I
Memory and Trauma

Chapter 1

WALKING MEMORY
Berlin's "Holocaust Trail"

Maria Pia Di Bella

In the Occidental world, Holocaust memorial museums are constructed following Washington, DC's United States Holocaust Memorial Museum (1993) assumption that, when it comes to the Holocaust, it is not enough for the visitors to learn its history; empathy or the identification with the victims are equally essential to their understanding. This has paved the way for a new type of museum: one not only able to display artifacts about the Holocaust but also able to address a difficult question: "How can one understand the places that speak about victims without being, herself or himself, a victim?"

Berlin has followed this pattern in the constructions of its museums (Jewish Museum in 2001 and Memorial to the Murdered Jews of Europe in 2005),[1] but has also developed—being aware of the expectations of the public at large and also of its own citizens—what I call a Holocaust trail where local sites of significant historical events leading to the Final Solution have been dedicated following multiple initiatives.[2]

Freedom without Walls

The fall of the Berlin wall (1989) was seen in the Occidental world as the end of the Cold War era and the beginning of a new, free one (*Freedom Without Walls*). The town itself became the symbol of this expected millennium. This event brought about for today's generation many changes, among which we find a different way of dealing with its past. In fact, today's generation looks at the history of its society as a heritage whose elements receive their moral and political significance through the judgment of the generation that follows—their own. In contrast, the pre-World War II generations tended to look at their past as a domineering tradition imposed on them by their elders, which they had to imitate or to contest. Whether they accepted or contested the tradition, it was seen as a dominant and present force.

Notes for this chapter begin on page 24.

Maria Pia Di Bella

I make my point by illustrating the way in which, since the early 1990s, the town of Berlin has developed a Holocaust trail. These street memorial sites came into existence thanks to multiple initiatives (the town's senate, the boroughs [*Bezirke*], the Jewish community, private citizens, artists like Gunter Demnig or Christian Boltanski, the League for Human Rights, etc.). Thus, since the late 1980s, the city of Berlin has become a true laboratory for the politics of memory concerning the crimes of the Nazi state and for the sufferings of the citizens that fell victim to the state's genocide.

Sites: Dates and Genres

In order to show the importance of Berlin street memorial sites as secular pilgrimage sites one should examine them from different angles. The first one should be the construction dates of these sites, which are important for the overall analysis. The first dedication was done in 1952 (the commemorative site of Plötzensee prison, the former Nazi execution site), and by 1990 only six more street memorials were added. But in 2005 Berlin had twenty-two dedicated memorials in different parts of the town.[3]

If we take 1990 as the dividing line, we can immediately notice that only five street memorial sites were constructed between 1945 and 1989, while eighteen more street memorial sites were added between 1990 and 2005, without counting museums that are not discussed in this article. The very slow progress, in the construction of these street memorials, until 1990, has many reasons, among which is the resistance of the institutions that were requested to take the responsibility for their actions in the Nazi period, and Germany's division in two different antagonistic states, with two different attitudes toward the Shoah.

Another way to examine Berlin's memorial sites in order to show their importance as a secular pilgrimage is to divide them by genre, whether they are tied to specific events, or related to the genocide of a whole community. I first focus on specific events that are retrospectively read as indications of things to come.

The perfect example is the underground memorial called The Empty Library—by Israeli artist Micha Ullman—on Bebelplatz, located at the spot where on 10 May 1933 national-socialist students burned twenty thousand books written by more than four hundred different authors who were considered to be "un-German." The empty white lit library bookcases in the subterranean room—covered with a glass roof in a shape of a window for passersby to look down into it—symbolize the loss of the twenty thousand books. A bronze plaque nearby bears a quote by Heinrich Heine (1797–1856): "Das war ein Vorspiel nur, dort wo man Bücher verbrennt, verbrennt man am Ende auch Menschen" (That was a prelude only. Where one burns books one will ultimately burn people

Illustration 1.1: Tourists around the Empty Library in Bebelplatz (Berlin, 2011). Photo: Maria Pia Di Bella

also). The memorial, set up at the initiative of the Berlin Senate, was inaugurated on the sixty-second anniversary of this infamous fire, on 10 May 1995.[4]

Next, specific events serve to reconstitute and remind us of the way in which the genocide was organized. These events cannot be separated from the type of locations in which they took place. I think in particular of "collective points of assembly" and of "train stations" for, as we all know, Jewish citizens were first summoned to come to a particular place and, subsequently, taken to a particular train station close by in order to be carried to death camps.

Assembling Points and Train Station in Moabit

Some of these places—the assembling points and the train stations—have become major memorials sites honoring the victims' memory, for example the Levetzow Deportation Memorial (corner Jagowstrasse), set up at the initiative of the Berlin Senate and dedicated in 1988. The Levetzow Deportation Memorial stands at the site of what used to be the Liberal Synagogue Levetzowstrasse (Moabit)—one of the largest synagogues in the city, used from 1941 on as a central collection point for Jews to be deported to death camps, destroyed during an air raid in 1944, and torn down in the mid-1950s. Up to one thousand Jews per night were taken from their homes and brought here, all in all thirty-seven thousand persons.

On this site now stands a sculpted ramp and a cattle train car, in front of which is a marble statue representing people huddled together, erected by two architects, Jürgen Wenzel and Theseus Bappert, as well as a sculptor, Peter Herbrich.[5] Behind the marble statue, a tall metal headstone rises with the deportation dates inscribed on it, the number of people carried off at each deportation, and where they were taken.

The memorial in Putlitzbrücke (Putlitz Bridge)—dedicated in 1987—recalls the deportation of thirty-two thousand Jewish citizens carried from the "collective points of assembly" at the Liberal Synagogue Levetzowstrasse. In fact, when deportations started, Jews were taken to the Grunewald Railway Station, but from March 1942 on to the nearby Moabit freight depot (or at Anhalter Bahnhof) across from the Putlitz Bridge, to cattle train cars on their way to the death camps.[6]

The Putlitz Bridge Deportation Memorial consists of a two-and-a-half-meter high (8 feet 4 inches) sculpture in stainless steel by Volkmar Haase, the front part separated from the back part, each part leaning in opposite directions, with the front part toward the onlooker, and the back part backward, as if the moment of separation between the two parts is taking place "here and now." The front part resembles a gravestone and the David Star on top of it is there to state that it is a gravestone to the thirty-two thousand Jews deported to extermination camps from this particular bridge. On its lower part one can read the following poem:

"Stufen, die keine Stufen mehr sind. Eine Treppe, die keine Treppe mehr ist. Abgebrochen, Symbol des Weges der kein Weg mehr war fuer die, die ueber Rampen, Gleise, Stufen und Treppen diesen letzen Weg gehen mussten."

(Steps which are no longer steps. A stair which no longer is a stair. Broken off, symbol of a journey which was no longer a journey for those who across ramps, tracks, steps and stairs had to go this last way.)

The back part on the contrary—which shoots up to the sky to fall immediately in what seems to be a flight of stairs—suggests the break of the relationship between Jews and Germans during the Nazi regime.[7]

It is important to link the places that were functionally linked, like the two examples discussed here. The historian has to do it in order to highlight the apparatus of genocide to show that this crime could not have been realized without the full support of the state bureaucratic and technological means. And the visitors or the pilgrims wishing to walk the Berlin Holocaust trail would have the option to access certain memorials in a sequel in order to understand—thanks to these memorials—that the unconceivable happened and how it happened.

Train Station in Grunewald: Platform 17 Memorial

A train station memorial that stands out, in my view, is Platform 17 (Gleis 17 Mahnmal) at Grunewald Railway Station, built to commemorate the deportation of all Berlin Jews, and dedicated on 27 January 1998.[8]

The role played by Deutsche Reichsbahn (German State Railway) during the National Socialist regime was evident to all historians: without the railway, at that time run by Deutsche Reichsbahn, the deportation of the European Jews to the extermination camps would not have been possible. But for many years, both the Bundesbahn in West Germany and the Reichsbahn in East Germany were unwilling to take a critical look at the role played by Deutsche Reichsbahn in the Nazi crimes.

When the reunified railways were merged to form Deutsche Bahn, the management board decided to erect one central memorial at Grunewald station on behalf of Deutsche Bahn in order to commemorate the deportation transports handled in all Berlin by Deutsche Reichsbahn between 1941 and 1945, and to keep the memory of the victims of National Socialism alive.

Thus the memorial Platform 17 was installed at the Grunewald train station's deportation track—which ceased to be in use—reachable by going up the stairs once in the station. The memorial is composed of 186 cast steel grates—each grate representing a transport that took place— assembled in chronological order and set in the ballast next to the platform edge. The design was created by a team of German architects, Nikolaus Hirsch, Wolfgang Lorch, and Andrea Wandel, of the Wandel Hoefer Lorch and Hirsch Office.[9] Each cast steel grate is engraved with a full date, number of Jews deported on that date, the place of departure (Berlin) and destination. Thus, the first grate reads: "18. 10. 1941 / 1251 Jews / Berlin - Lodz." And the last grate reads: "27. 3. 1945 / 18 Jews / Berlin - Theresienstadt."

The memorial is steps away from the monument created by Karol Broniatowski in 1991, a concrete block embedded with several human silhouettes—five silhouettes are distinctly visible, two are more faded—inserted in the passage taken by Berlin Jews on the way to the rail track 17 for deportation. Platform 17 memorial (or double memorial) has become one of the most popular memorial in Berlin, being part of many touristic tours of the town.

Artists on Grosse Hamburger Strasse

Berlin's memorial sites division by genre has also to take into account the artists' contribution. Behind every memorial in Berlin's streets there is an artist that came up with an original idea allowing her or him to win the contest promoted by the town's senate. Or the artist came up with the idea and proposed it to the town. The most important difference between the works is that some—the majority in fact—had to be done on the spot while others could be done anywhere, possibly allowing the artist a greater freedom to express herself or himself. But the works artists presented in order to honor the memory of the Shoah are generally outstanding and they contribute to the public knowledge of Berlin's past history and to the spectators' empathy with the victims.

One example is the trail one can walk in Berlin Mitte's Grosse Hamburger Strasse. This trail goes north from Christian Boltanski's The Missing House (1990) to The Abandoned Room (1996) in Koppenplatz (Koppen Square). Steps away from these two postwar contributions is a plaque—dedicated in 1985—next to the Jewish cemetery entrance (Grosse Hamburger Strasse 26). It sits on the site of the former Jewish Home for Elders emptied in 1942 by the Gestapo and used as a "collective point of assembly," commemorating the more than fifty-five thousand Berlin Jews deported, and a sculpture, designed by Will and Mark Lammert (1985).[10]

The Jewish cemetery itself, the oldest in Berlin (1672), had reached its capacity in 1820. In 1943 the Gestapo ordered all graves to be destroyed. In 2009, the gravestones were returned to the cemetery; even earlier, in 1990, a reproduction of the original tomb of Moses Mendelssohn was erected once again in the cemetery. The remodeling of the cemetery was financed by the Berlin Senate and the Jewish community.[11]

After the fall of the Berlin wall, the reunification of the two Germanies was celebrated with the exhibition Die Endlichkeit der Freiheit (1990), where the French artist Christian Boltanski was invited to participate. He and his students found—on Grosse Hamburger Strasse 15–16—the empty space, formerly occupied by a building destroyed in February 1945 during an aerial bombardment. And they discovered that, among its residents, twenty-eight Jews had been sent to death camps. He decided to use the empty space between the walls to construct a memorial dedicated to "absence," which he called The Missing House. Plaques were set up approximately where the residents used to live, with their full names, dates of birth and death, and their profession. "Plaques indicate the approximate space occupied by Jewish and non-Jewish residents, testifying to a diversity that was lost with Nazi decrees against the Jews and removal of the Jewish population from Berlin."[12]

North of Grosse Hamburger Strasse, in Koppenplatz—at the far end of a peaceful public garden where adults come to read or chat and children to play—The Abandoned Room (1996) created by Karl Biedermann and Eva Butzmann. This sculpture represents the sudden departure of Jewish citizens from their homes: on a rather large rectangular parquet floor is a rectangular brown bronze table surrounded by two brown bronze chairs, one of which is with its back on the floor, suggesting a violent confrontation or a swift departure. The four sides of the rectangular floor are framed with an excerpt from a poem by Nelly Sachs (1891–1970), a Jewish German poet.[13]

Places of Remembrance in Schöneberg

The Schöneberg borough has always been appreciated by Jewish intellectuals who came to live in its peaceful streets and handsome Wilhelmine-style houses from the turn of the century on—Gisèle Freund (1908–1933), Albert Einstein (1918–1932), Billy Wilder (1927–1928), Wilhelm Reich (until 1933), and Claudio Arrau (1930–1937). Today the streets where they lived are covered with plaques reminding visitor and passerby of their presence (Blakenburg 2011: 45-49).

Thus when the senate and the Schöneberg borough had to choose a memorial to the Jews in "the awakening spirit of working through the past," (Wiedmer 2009: 7) they dared to choose a fascinating project by Renata Stih and Frieder Schnock that not only reminds the onlookers of the past, but also informs them of the many official Nazi laws or regulations that kept coming out regularly—from 1933 on—and commenting on each law via an image or a symbol.[14]

The result is not simply a Holocaust trail, but one that adds a "labyrinth spirit" that the two artists use in order to remind the onlookers of the recent past intricate history and, at the same time, to show all the absurd, comic, and murderous implications of these Nazi laws. Their project, called "Places of remembrance: Isolation and deprivation of rights, expulsion, deportation and murder of Berlin Jews in the years 1933 to 1945," is made up of eighty signs, each one bolted rather high to lampposts. On one side of the sign there is a text that gives the gist of an anti-Jewish Nazi regulation or law with its date, while on the other side there is a "comment" by Stih and Schnock on the specific regulation or law via an artistic image or symbol. The eighty signs are scattered all around the Bayerisches Platz area. To help visitors to find their way around, individual maps are available on request and two general ones are posted in the square. The visit may take two hours or several days, depending on the way one wishes to progress in this labyrinthical trail.

Stolpersteine (Stumbling Stones)

Since 1992, the Berlin sculptor Gunter Demnig (born in 1947) has been producing handmade brass memorial cubes, which he calls *Stolpersteine* (stumbling stones). Official approval in Germany of his project came in 2000 (City of Cologne 2007: 14-15). The cubes are called stumbling stones because one unexpectedly trips over them—figuratively speaking—while strolling through the city sidewalks.[15] They are brass cobblestones engraved in the center with, on the first line, the words HIER WOHNTE (here lived) followed, in the next line, by the person's full name in capital letters, with the maiden name for married women in the third line, coming after the word GEB. (name of birth); the fourth line has JG (for Jahrgang, year of birth); the fifth line has generally the word DEPORTIERT (deported) followed by the full date of arrest; next line, the name of the concentration camp; on the following line the word ERMORDET (murdered) with the date of assassination and, on the last line, the place of assassination, often AUSCHWITZ.

Each *Stolperstein* is placed in front of a house where a person lived last and from where he or she was deported by the Nazis, literally pointing—in Berlin's streets—to the number of victims sent to their death that nobody could have ignored. *Stolpersteine* are the perfect answer to the people who say that they never noticed that their Jewish neighbors were taken away.[16]

Illustration 1.2: Gunter Demnig discussing with the people who ordered the Stolpersteine where to put the Stolpersteine. Photo: Maria Pia Di Bella

Illustration 1.3: Gunter Demnig and his aide working in order to insert the six Stolpersteine in the macadam. Photo: Maria Pia Di Bella

Illustration 1.4: The six Stolpersteine inserted in the macadam surrounded by flowers. Photo: Maria Pia Di Bella.

Maria Pia Di Bella

Demnig says that he wants to bring back—with his art—the names of the millions of Jews, gays, resistance fighters, and Gypsies who perished at the hands of the Nazis between 1933 and 1945. In fact this is the true significance of the *Stolpersteine*: they are an individual tribute and not a collective one, they allow the name of the persons to survive as long as the stumbling stones are kept in the pavement of cities that witnessed the murder of millions of people before and during World War II.

It is an arduous task to find a Holocaust trail by following Demnig's *Stolpersteine* in Berlin. Wikimedia does provide a listing on Berlin's *Stolpersteine*.[17] Most likely in the future specific lists pointing to possible trails will exist and will be used by persons eager to see *de visu* the *Stolpersteine* of people they loved or knew about.

Politics of Memory

The Berlin Senate has played, since the 1990s, a very active role in the politics of memory, creating a nonmuseal form of memory that makes the architecture and the space of the city the incorporation and the carrier of memory: the Gleis 17 in the still fully functioning Grunewald train station, the Levetzow Synagogue Memorial at the site of a much used playground, and the memorial on the Putlitz Bridge Deportation Memorial—often attacked by right-wing groups, and always renovated—are important traffic point, all of which can be easily accessed by foot.

Illustration 1.5: Steglitz, the Spiegelwand. Photo: Maria Pia Di Bella.

Illustration 1.6: Berlin-Grunewalld, "Gleis 17." 18-10-1941: First train from Berlin to Lodz, carrying 1251 Jews. Photo: Maria Pia Di Bella.

Illustration 1.7: : Berlin-Grunewalld, "Gleis 17." 27-3-1945: Last train going from Berlin to Theresienstadt with 18 Jews on board. Photo: Maria Pia Di Bella.

Maria Pia Di Bella

Is Berlin's politics of memory successful? During my visits (2010, 2011, 2012), I did not notice too many visitors in the places I went, except at Bebelplatz, around The Empty Library memorial, and at Grunewald's station Gleis 17 Memorial. But traces of motivated visitors are always perceptible due to the fact that they leave behind small stones or a rose as a religious or familial duty.[18] And during discussions with friends and colleagues on Berlin's development, I realized the impact these memorials have on them as well, especially the *Stolpersteine*.

For the time being tourists seem to prefer official well-known historical sites, like the Reichstag, restored during the 1990s as the center of the German Federal Republic government, when the capital was moved from Bonn back to Berlin, now the symbol of the new Germany. Or like the House of the Wannsee-Conference, the villa that on 20 January 1942 harbored the discussion, organization, and implementation of the decision to deport and murder European Jews in East Europe.[19] But once the trail is fully completed—and it will take some years to be completed—Berlin's Holocaust trail will be a major attraction as a secular pilgrimage for tourists and local residents alike.

In conclusion I want to underline that the memorials and the Holocaust trail do not exist solely to relate a history or to allow people to empathize but also to address the memory and the conscience of the heirs of those who planned and executed the Holocaust. These memorials represent an architecture that condemns a whole period of German history and make it impossible to forget it. This also explains why it took many decades to build the memory of genocide in lasting monuments all over the city.

Maria Pia Di Bella is co-founder and co-editor of the journal *Journeys: The International Journal of Travel and Travel Writing*. She is a senior research scholar at IRIS-EHESS, Paris, and Research Affiliate at the Harvard Divinity School. She has published monographs on popular religions and cultures of punishment and penitence, including *La Pura verità. Discarichi di coscienza intesi dai Bianchi* (1999), *Dire ou taire en Sicile* (2008), and *Essai sur les supplices. L'état de victime* (2011). She has edited, among others, *Vols et sanctions en Méditerranée* (1998) and with James Elkins, *Representations of Pain in Art and Visual Culture* (2012). Since 2011 she works on street memorials in Berlin concerning the victims of genocide and the function of these memorials as "symbolic reparations."

Notes

1. See Bernau, 2005; Foundation for the Memorial to the Murdered Jews of Europe, 2005; Rauterberg, 2005; Schlör, 2005.
2. I suggest in this article the appellation of the Holocaust trail following the example of Boston's Freedom Trail. In March 1951, Bill Schofield—columnist and editor for the

Herald Traveler—wrote to suggest that Boston citizens get together to create the link that would "tie the story of the American Revolution together" from ... "Paul Revere's house and the Old North Church to the Old State House and the Old South Meetinghouse." ... A walking trail was designated on Boston's sidewalks in front of sixteen historically significant buildings and locations. The new path was called Freedom Trail (http://www.thefreedomtrail.org/freedom-trail/establishing-ft.shtml [accessed 3 September 2012]). I am aware that in Berlin the distance from one site to the other is sometimes very long and that I should possibly use the plural "trails" in order to convey a better idea of what we find on the ground. But until 1942 Jewish citizens were taken by foot from the Liberal Synagogue Levetzowstrasse (Moabit) to the far away Grunewald Railway station, instead of being taken to the nearest stations—Moabit freight depot or Anhalter Bahnhof. Thus these distant places were linked historically and this is the reason why I use singular "trail."

3. In this article I am not able to further discuss the topic of the trail's construction dates but I still want to give an overview of the phenomenon by listing the dedicated sites in Berlin. In 1952, the execution shed at Plötzensee prison was turned into a memorial operated by the Memorial to the German Resistance institution to commemorate the 2,891 people executed there by the Nazis; in 1963, a plaque and a monument was erected to the destroyed Münchener Strasse Synagogue, in Schöneberg (built in 1909 and destroyed in 1956); in 1967, a steel sign—Places of Terror We Must Never Forget—was erected at the Wittenberg Platz, at the initiative of the League for Human Rights; in 1985, a memorial stone was erected: it commemorates the "collection point" organized by the Gestapo in 1942 at Grosse Hamburger Strasse 26, from where more than fifty-five thousand Berlin Jews were deported; in 1987, the Putlitz Bridge Deportation Memorial was erected in Moabit; in 1988, the Levetzow Synagogue Memorial was erected in Moabit; on December 1989, the restoration of the Jewish Congregation (*Adass Yisroel*) was ordered and the reconstruction of the Congregation began. The new synagogue was consecrated in 1990 in the ancestral community center and Torah scrolls were brought over from Israel; in 1990, The Missing House was dedicated in Grosse-Hamburger Strasse (Mitte); during the 1990s, the Reichstag was restored as the symbol of the new Germany. The new glass dome by Norman Foster speaks of the idea of "transparency" in government. In its cellar, Christian Boltanski installed an archive related to past and present members of the German parliament. Outside the Reichstag itself, we find a Memorial to the Murdered Parliament Members from 1933 to 1945, a monument to those killed in concentration camps during the Nazi era, a Memorial to East Germans Killed Fleeing over Berlin Wall and graffiti from Soviet soldiers; in 1992, the Memorial and Educational Site House of the Wannsee Conference was opened, on the fiftieth anniversary of the meeting that established Nazi policy on the "final solution of the Jewish question"; in 1993, Places of Remembrance: Isolation and Deprivation of Rights, Expulsion, Deportation and Murder of Berlin Jews in the Years 1933 to 1945 was erected in the Bayerisches Platz area (Schöneberg); in 1993, the former Neue Wache (New Guard House) in Unter den Linden became the Central Memorial of the Federal Republic of Germany for the Victims of War and Tyranny. The interior, reconstructed to look the way it did in 1931, contains a replica of the sculpture *Mother with Dead Son* by Käthe Kollwitz; in 1995, Der Spiegelwand (Mirrored Wall Memorial)—a memorial for the former Synagogue Haus Wofenstein on Düppelstrasse, designed by architects Wolfgang Göschel and Joachim von Rosenberg, in cooperation with Berlin historian Hans-Norbert Burkert, was erected on Hermann-Ehlers-Platz (Steglitz). It consists of eighteen highly polished stainless steel plates—thirty feet wide and eleven feet high—which have engraved the names, birthdates, and addresses of 1,723 deported Jews, including 229 from Steglitz, arranged by transport in Nazi concentration camps. At the same time, the memorial reflects the onlookers and everyday events on the square; in 1995, The Empty Library was erected in Bebelplatz (Mitte); also in 1995, Block of Women, an ensemble of protesting and mourning women monument by Ingeborg Hunzinger was erected on the site of the Rosenstrasse

Protest (1 March 1943); in 1996, The Abandoned Room was erected in Koppenplatz (Mitte); on 27 January 1998, a memorial to the memory of the deportation of all Berlin Jews from rapid transit stations was dedicated in the historic Grunewald platform number 17; also in 1998, JUST STOP! The Bus Shelter Project was erected at Kurfürstenstrasse 115/116, at the initiative of Ronnie Golz; in November 2001, a memorial was dedicated to Rabbi Menchem M. Schneerson at the home where the Rabbi lived (Hansa Ufer 7); on 8 May 2005, the Memorial for the Murdered Jews of Europe—a field of columns designed by the American architect Peter Eisenman near the Brandenburg gate—opened; on 27 May 2008, The Memorial to Homosexuals Persecuted under Nazism in Berlin designed by artists Michael Elmgreen and Ingar Dragset, was inaugurated in the Tiergarten, next to the Reichstag and the Brandenburg Gate; also in 2008, The Silent Heroes Memorial Center—financed with funds from the German government and the Klassenlotterie Foundation Berlin—was opened in the Haus Schwarzkopf building in the Rosenthaler Strasse in Berlin's Mitte borough; on 6 May 2010, the documentation center Topography of Terror—designed by the architect Ursula Wilms and the landscape architect Heinz W. Hallmann—located on the grounds of the former Prinz Albrecht Palais, on Wilhelmstrasse, where the Gestapo and the Security Service of the SS had their headquarters, opened: their exhibitions examine the Third Reich's systematic persecutions and repressions. Remains of former buildings bear witness to the history of the site; on 24 October 2012, a memorial site to the memory of the murdered Sinti and Roma, designed by the Israeli artists Dani Karavan—a circular pool of water with a small plinth in the middle where a fresh flower will be laid every day—was unveiled in the Tiergarten.
4. Quoted from: www.chabadberlin.de, and Wikipedia. Accessed on August 30, 2012.
5. Quoted from http://fcit.usf.edu/holocaust. Accessed on August 30, 2012.
6. See *Berlin Anhalter Bahnhof*, http://en.wikipedia.org/wiki/Berlin_Anhalter_Bahnhof. Accessed on August 29, 2012.
7. See also http://de.wikipedia.org/wiki/Deportationsmahnmal_Putlitzbrücke and http://memorialmuseums.eu/eng/staettens/view/1402/Deportation-Memorial-on-Putlitzbr%C3%BCcke. Accessed on August 29, 2012.
8. Quoted from: http://www.deutschebahn.com/site/bahn/en/group/history/topics/platform17/.html. Accessed on August 30, 2012.
9. http://www.deutschebahn.com/en/group/history/topics/platform17_memorial.html. Accessed on August 29, 2012.
10. At Grosse-Hamburger-Strasse 27 there is also a reminder of the boys' school of the Jewish community in an inscription above the portal. See, for information on Grosse-Hamburger-Strasse, Rebiger, 2010: 9-12; see also http://www.memorialmuseums.org/eng/staettens/view/1422/Jewish-Sites-of-Remembrance-in-Gro%C3%9Fe-Hamburger-Stra%C3%9Fe, and also the Center for Holocaust & Genocide Studies, University of Minnesota, http://www.chgs.umn.edu/museum/memorials/berlin/index.html. Accessed last on August 27, 2012.
11. Quoted from: http://memorialmuseums.eu/eng/denkmaeler/view/1422/Jewish-Sites-of-Remembrance-in-Grosse-Hamburger-Strasse. Accessed on August 29, 2012.
12. Quoted from: http://www.chgs.umn.edu/museum/memorials/berlin. See also http://artspla.over-blog.com/article-la-memoire-the-missing-house-de-christian-boltanski-74814976.html. Accessed on August 30, 2012.
13. The full quotation in Koppenplatz goes: "[1] ... O die Wohnungen des Todes,/ Einladend hergerichtet/ Für den Wirt des Hauses, der sonst [2] Gast war./ O ihr Finger,/ Die Eingangsschwelle legend/ Wie ein Messer [3] Zwischen Leben und Tod-/O ihr Schornsteine,/ O ihr Finger,/ Und Israels Leib im Rauch durch [4] Die Luft!" Nelly Sachs (10. Dezember 1981 Berlin- 12. Mai 1970 Stockholm). Nelly Sachs was awarded the Nobel Prize in Literature in 1966.
14. Stih's and Schnock's Places of Remembrance is an "art in public space" project, approved by the Senate of Berlin (Die Senatsverwaltung für Bau-und Wohnungswesen),

which was in charge of all art in public space projects at the time the project was conceived.
15. See Kirsten Grieshaber, "German Artist Gunter Demnig Revives Names of Holocaust Victims," at http://www.artdaily.com/index.asp?int_sec=2&int_new=39124. Accessed on Sept. 2, 2012. See also "List of cities by country that have *Stolpersteine*," at http://en.wikipedia.org/wiki. Accessed on Sept. 2, 2012.
16. Rosetta Loy (2000) recounts in a moving way her childhood souvenirs when, in Rome, her Jewish neighbors started to "fade away" or to be taken away by the Fascist police.
17. For the complete alphabetical list of names found in Berlin's *Stolpersteine*, see http://commons.wikimedia.org/wiki/User:OTFW/Bilder/Gedenktafeln/Stolpersteine. According to this site, the number of *Stolpersteine* in Berlin today is 2,950. Accessed on Sept. 2, 2012.
18. "Even when visiting Jewish graves of someone that the visitor never knew, the custom is to place a small stone on the grave using the left hand. This shows that someone visited the gravesite, and is also a way of participating in the mitzvah of burial. [...] Another reason for leaving stones is to tend the grave. In Biblical times, gravestones were not used; graves were marked with mounds of stones (a kind of cairn), so by placing (or replacing) them, one perpetuated the existence of the site," cited from http://en.wikipedia.org/wiki/Bereavement_in_Judaism. Accessed on Sept. 3, 2012.
19. For information on Berlin's memorials, go to the Information Portal to European Sites of Remembrance, a project of the Foundation Memorial to the Murdered Jews of Europe. It is part of the exhibition of the Information Centre under the Field of Stelea of the Holocaust Memorial in Berlin, see: http://memorialmuseums.eu/pages/home. Accessed last on August 30, 2012.

References

Bernau, Nicholas. 2005. *Holocaust Memorial Berlin*. Berlin: Stadtwandel Verlag Daniel Fuhrhop.
Blakenburg, Gudrun. 2011. *Das Bayerische Viertel in Berlin-Schöneberg. Leben in einem Geschichtsbuch*. Berlin: Hendrik Bässler Verlag.
City of Cologne's Documentation Center. 2007. *Stolpersteine: Gunter Demnig and his Project*. Cologne: Emons.
Foundation for the Memorial to the Murdered Jews of Europe. 2005. *Materials on the Memorial to the Murdered Jews of Europe*. Berlin: Nicolai.
Loy, Rosetta. 2000. *First Words: A Childhood in Fascist Italy* (La Parola ebreo, 1997). New York: Metropolitan Books.
Rauterberg, Hanno. 2005. *Holocaust Memorial Berlin, Eisemann Architects*. Baden: Lars Muller.
Schlör, Joachim. 2005. *Memorial to the Murdered Jews in Europe Berlin*. Munchen: Prestel.
Rebiger, Bill. 2010. *Jewish Sites in Berlin*. Berlin: Jaron Verlag.
Wiedmer, Caroline. 2009. "Remembrance in Schöneberg." Pp. 7-10 in *Places of Remembrance. Isolation and Deprivation Of Rights, Expulsion, Deportation and Murder of Berlin Jews in the Years 1933 to 1945. Memorial in Berlin-Schöneberg.* Ed. Renata Stih and Frieder Schnock. Translations, Images and Texts, VG BildKunst. Bonn-Berlin-New York City: ARS.

Chapter 2

TOURING THE AFRICAN DIASPORA

Cheryl Finley

> There is no place you or I can go, to think about or not think about, to summon the presences of, or recollect the absences of slaves. There is no suitable memorial, or plaque, or wreath, or wall, or park, or skyscraper lobby. There's no 300-foot tower, there's no small bench by the road. There is not even a tree scored, an initial that I can visit or you can visit in Charleston or Savannah or New York or Providence or better still on the banks of the Mississippi. And because such a place doesn't exist ... the book [*Beloved*] had to.
> —Toni Morrison (1989)

In 2007, at the twelfth installment of *Documenta*, the international contemporary art fair held every five years in Kassel Germany, an artist from Benin—Romuald Hazoumè—was awarded the coveted Arnold Bode Prize for the first time for his installation *Dream*. Suspended from the ceiling of the Orangerie's "Aue Pavillon," a greenhouse-like structure built to protect artworks from the rain, *Dream* resembled a large, long pirogue, of the type often seen in picturesque West African fishing villages. Hazoumè's boat was constructed of 421 painted petroleum canisters, each one representing an economic and political refugee, who has paid dearly for his or her place on the boat, but whose safe passage from Africa to Europe was not at all guaranteed. These plastic petrol canisters, used to transport smuggled fuel from Nigeria to Benin, are emblems in perpetual motion that permeate everyday Beninese life, as objects of survival, ritual, and even death. In Hazoumè's installation, they line up, some tagged in fluorescent orange spray paint with the word "Dream," before a seductive panoramic photograph of paradisiacal golden sands, palm trees, sea, crowds of eager children—those, as the inscription reads, are "damned if they leave and damned if they stay: better, at least, to have gone, and be doomed in the boat of their dreams."

Now characteristic of Hazoumè's mnemonic aesthetic, the plastic petroleum cans were also the building blocks of the sophisticated conceptual installation *La Bouche du Roi* (or the Mouth of the King), constructed between

---

Notes for this chapter can be found on page 34.

1997 and 2005, purchased by the British Museum in 2007, and now on a continuous, global touring schedule. In this work, arranged in the shape of the familiar late eighteenth-century slave ship icon circulated by proponents of abolition, they are transformed into masks representing African captives taken as slaves from the mouth of the River Porto Novo and their African and European captors. Yet in others, they feature in photographs of young Beninese petrol-carriers so shocking that their bodies are frequently eclipsed if not transformed into larger-than-life size bulbous, Michelin-like explosive figures, or in outdoor sculptures like *Serpent,* commissioned for the Victoria and Albert Museum's *Uncomfortable Truths* exhibition in 2007, or as playful, often witty contemporary puns on the tradition of African mask making, such as *Dogonne* and *Citoyenne.*

Touring the African Diaspora

The idea for *Touring the African Diaspora* is inspired in part by my new interdisciplinary project, *Re-Imagining the Grand Tour: Routes of Contemporary African Diaspora Art*, which weaves together several strands of my scholarly practice to examine the relationship among art, technology, global economies, museums, and tourism. The word "re-imagining" in the title of my project implies the constant invention of ideas and practices as well as flows of information and technology that reveal links to slavery, history, activism, economics, and aesthetics. But the word "touring" is not about destiny, nor is it a destination. Rather, perhaps, it is the recognition and understanding of "perpetual motion," of being in flux, that is the symbolic destination here. *Touring the Diaspora* recognizes that people seek experiences and outlets that enable them to cling to, to hold dear, those things that are familiar to them, while enabling them to identify with like communities in the midst of ground-shaking social, technological, economic, and political changes. This article aims to show how art, performance and technology are central to identity formation through an examination of mnemonic aesthetics and practices. One of the central questions it asks is: What roles do public memorials play in marking space, asserting identity, and claiming history diasporically, and what are some of the aesthetic challenges to creating public memorials in the age of hyper-memorialization?

Toni Morrison's prescient novel *Beloved* (1987) became a symbolic memorial to the history of slavery; its epigraph read, "Sixty Million and more," referring to the estimated number of African captives who died in the slave trade.[1] Her novel inspired similar literary works and fueled a boom in art exhibitions, memorials, and walking tours that unearthed and recognized the history and legacy of slavery in the United States and elsewhere in African diasporic communities. Twenty years after the publication of *Beloved*, Hazoumè's *La Bouche du Roi* became a national memorial to Britain's role in first promoting and then ending the slave trade. Though not a "plaque, or wreath, or wall, or park, or skyscraper lobby," to borrow from Morrison, *La Bouche du Roi*'s arrival at memorial status provides an apt example of a twenty-first century heritage trail created through an orchestrated act of national commemoration and a touring exhibition.

I want to chart these points of translation in black Atlantic art, geographically, conceptually, and historically. Hazoumè's little dreamboat provides an able vessel in which to cast off on our journey. For the metaphor of the dream, that is, its unattainability and fantasy-like nature is what makes dreaming possible, that artists can imagine new spaces of creation and new ideas. Touring also suggests a certain kind of aspiration and the possibility of movement into these spaces, of reclaiming, if not claiming spaces that were hitherto off limits, of mapping and walking new paths.

The multimedia installation *La Bouche du Roi* is named after a well-known site of memory on the coast of Benin from which African captives were transported during the transatlantic slave trade with the blessing of the king and the collusion of African, European, and American traders. The artwork comprises 304 plastic petroleum canisters made to resemble "masks" with the spout serving as a mouth and the handle as the nose. These dark plastic, bulbous canisters are arranged in rows forming the shape of the now iconic, schematic engraving of a slave ship successfully deployed by British abolitionists in the late eighteenth century to shed light on the horrors of the slave trade in order to build parliamentary support for its cessation in 1807. In Hazoumè's installation, each mask represents a living person with a name, a voice, and individual beliefs while hidden microphones whisper their presence in Yoruba. Painted symbols and small objects affixed to the canisters, such as "ibeji," literally translated to mean "twins," or the carved wooden figures representing (memorializing) the soul of a dead twin, reiterate Yoruba beliefs and the "orishas" or spirits to whom the enslaved might have prayed. Strategically placed between the masks following the pattern of the schematic template are cowry shells, rifles, tobacco, beads, spices, and liquor bottles, which reference the trade goods that would have been used to barter for human beings.

In an interview with the artist in March 2007 at the October Gallery in London, Hazoumè told me that he had seen pictures of *Description of a Slave Ship* (1789)[2]—the abolitionist print after which his installation is modeled—in textbooks and tourism brochures. It was a familiar image, he said, in certain parts of the Republic of Benin, including Porto Novo, where he lives, owing to the burgeoning commerce in heritage tourism and the UNESCO-sponsored slave routes project, both commercial ventures that have increased awareness about the shores of the Republic of Benin as important sites of memory of the transatlantic slave trade.

Providing a contemporary context for Hazoumè's slave ship is a video narrated by the artist discussing the treacherous, illegal gasoline trade from Nigeria to Benin.³ Still photographs from the video, contrasting individual motorcyclists weighed down by hazardous fuel canisters with paradisiacal beaches of the mouth of the River Porto Novo, illustrate the devastating, lingering effects of centuries of economic, social, and racial oppression. The darkened exhibition space creates an ominous yet somber mood for the viewer, providing a place of reflection and memorial.

La Bouche du Roi was acquired by the British Museum to commemorate the 2007 Bicentenary of the Parliamentary Abolition of the Transatlantic Slave Trade Act by the United Kingdom. The installation was on view at the British Museum in London from 22 March to 13 May 2007, prior to being sent on a carefully orchestrated tour to Hull, Liverpool, Bristol, Newcastle, and the Horniman Museum in South London.⁴ That large-scale installation used *Description of a Slave Ship* to enter into a conversation between the past of the slave trade and the present day corruption and loss of life that takes place along the route of the illegal gasoline trade between Nigeria and Benin.

During the bicentenary commemoration of the abolition of the slave trade by the United Kingdom, *Description of a Slave Ship* was at the center of several national and international commemorative events. It was featured in print form in numerous exhibitions around the country and even made part of the graphic imagery of postage stamps issued by the Royal Mail to mark the bicentenary. Indeed the endeavors to enable this image to stand as a marker for the memory of the Middle Passage and the graphic violence of the transatlantic slave trade against black and white bodies vary in scale, scope, and intent. In recent years, several attempts have been made to revisit *Description of a Slave Ship*, to embody it, to reinvent it, to memorialize it.

At Durham University in the United Kingdom, 274 school-age children commemorated the Bicentenary of the Abolition of the Slave Trade Act on 13 July 2007 by participating in a reenactment of *Description of a Slave Ship* on Palace Green in Durham City between the Castle and the Cathedral. Wearing black trousers and red T-shirts printed with rows of black figures, they placed their bodies on a full-size printout of the most recognizable bullet-shaped section of *Description of a Slave Ship*, to "have a sense of what the stifling space was like in the hold, to know the sufferings of African captives in the confines of dank slave ships."⁵

Cheryl Finley

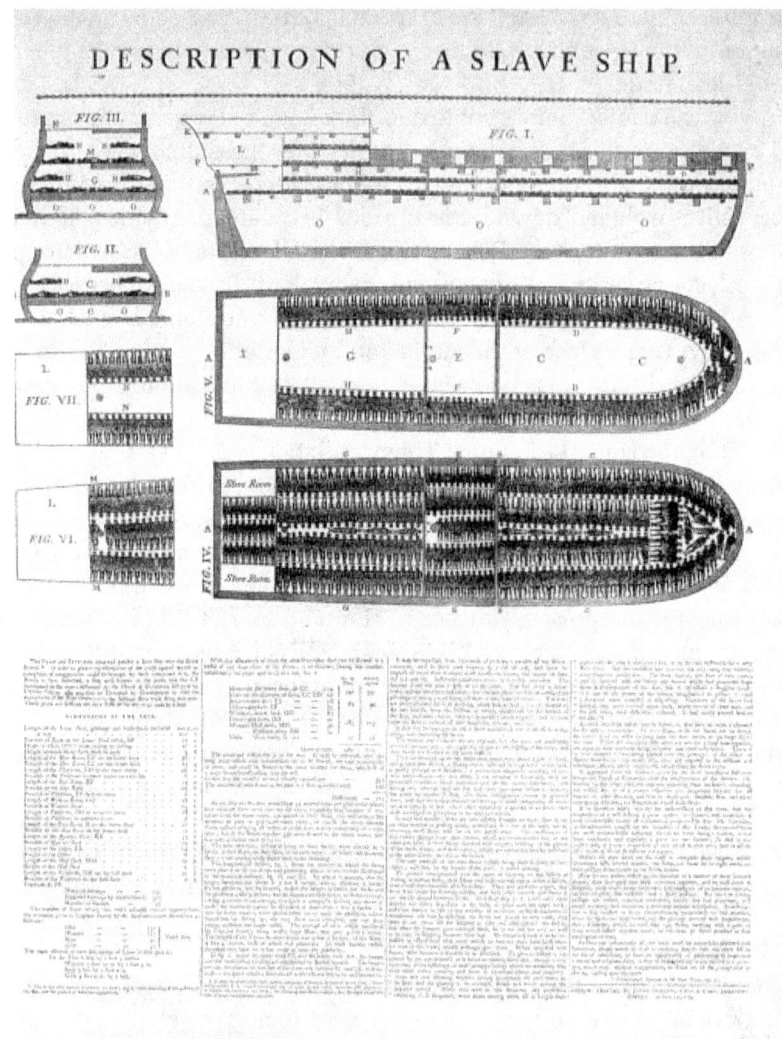

Illustration 2.1: *Description of a Slave Ship,* London Committee of the Society for the Abolition of the Slave Trade, engraving by James A. Phillips, 1789.

Aerial photographs taken of the scene recorded the life-size public performance of this unforgettable image while reinforcing the exacting technology of vision that has structured the visual empathy of *Description of a Slave Ship*, seeking to emulate the same way of seeing that the image has demanded for more than two hundred years. Not unlike the touring presentation of

Hazoumè's *La Bouche du Roi*, this was one of many times the public performance of *Description of a Slave Ship* was reenacted in the United Kingdom that year, revealing a practice of mnemonic aesthetics underlying contemporary presentations of the art of slavery.

The 2007 Bicentenary and the Practice of Mnemonic Aesthetics

The tour of Romuald Hazoumè's *La Bouche du Roi* to six significant venues around the United Kingdom from 22 March 2007 to 1 March 2009 charts a carefully planned program of national absolution, repentance, and perhaps, admission of guilt, but it also demonstrates the practice of mnemonic aesthetics. All venues had ties to the history of the slave trade or its abolition. London was a principal port engaged in the slave trade in the late seventeenth and early eighteenth century. It also profited heavily from slave-grown sugar after the abolition of the slave trade. Liverpool was the leading slave trading port in all of the United Kingdom, especially in the last half of the eighteenth century and *La Bouche du Roi*'s stop in that city coincided with the opening of the new International Slavery Museum at the Albert Dock. Bristol was an important slave trading port. Manchester manufactured trade goods that were used to barter for African captives. Hull was the home of MP William Wilberforce, who successfully argued before Parliament for the abolition of the slave trade. In conjunction with the installation of *La Bouche du Roi* in that city, the newly renovated Wilberforce Institute for the Study of Slavery and Emancipation was celebrated.

La Bouche du Roi's choreographed movements around Britain suggested more of a march, a procession, a pilgrimage of reconciliation, perhaps as a way of forging a new national memory and charting a symbolic heritage trail. The repetition of bodies and sections that are illustrated in the slave ship icon are not only reworked in the mask-like plastic petroleum canisters that comprise *La Bouche du Roi* but a mnemonic aesthetic was also apparent in the very circulation of the installation to its various venues.

La Bouche du Roi's tour, moreover, emulates the vast dissemination of the original prints of *Description of a Slave Ship* to the countryside, city centers, and abroad during the busiest period of abolitionism from 1789 to 1807. Through this state-led effort, *La Bouche du Roi*, and the historical and contemporary references it makes, was promoted as a symbol of national memory and reconciliation during the bicentenary celebrations.

Cheryl Finley

Cheryl Finley is the Distinguished Visiting Director of the Atlanta University Center Art History + Curatorial Studies Collective. She is a curator, contemporary art critic and award-winning author noted for *Committed to Memory: the Art of the Slave Ship Icon* (Princeton UP, 2018) and *My Soul Has Grown Deep: Black Art from the American South* (Yale University Press, 2018). On leave from Cornell University, where she is Associate Professor of Art History, Dr. Finley's research examines the global art economy, focusing on the relationship among artists, museums, biennials and migration in the book project, *Black Market: Inside the Art World*.

Notes

1. Toni Morrison, *Beloved* (New York: Random House), 1987.
2. "219 Years Ago: Description of a Slave Ship," http://blogs.princeton.edu/rarebooks/2008/05/ship_brooks.html (accessed 18 August 2012). Published in 1789, "Description of a Slave Ship" attained an iconic status in the antislavery movement in the Kingdom of Great Britain and in the United States. It provided the first graphic illustration of how tightly Africans were packed as human cargo on slave ships. Mathematical calculations describing the space allotted to each man, woman, boy and girl and documentations of incidents of torture accompanied the image. By the end of the eighteenth century, more than 200,000 impressions of the print had been reproduced and disseminated around the Atlantic regions. I explore this aspect of black Atlantic history in *Committed to Memory: The Slave Ship Icon in the Black Atlantic Imagination* (Princeton University Press, 2018).
3. In Hazoumè's interview by BBC Africa, he discusses Benin's illegal gas smugglers who precariously ferry contraband gas between Nigeria as well as their local customers, see http://www.youtube.com/watch?v=0L4gejIY2dk (accessed 19 August 2012).
4. *La Bouche du Roi* was exhibited previously at the Menil Collection, Houston (2005) and the Musée du Quai Branly, Paris (2006). See "Setting up 'La Bouche du Roi.' An Art Work by Romuald Hazoumè at the British Museum," pictures and edit Alistair MacKillop, 2007, at http://www.youtube.com/watch?v=rCN0sic-jI8 (accessed 7 August 2012).
5. "Palace Green Transformed into a Slave Ship," *Durham First*, Winter 2007, http://www.dur.ac.uk/durham.first/winter07/slaveship/ (accessed 2 August 2012).

Chapter 3

A WARTIME CINEMATIC RECREATION OF THE JOURNEY LINKING CHINA AND JAPAN IN THE MODERN ERA

Joshua A. Fogel

One of the most famous voyages in the modern history of East Asia occurred in 1862 when, for the first time in over three centuries, Japanese were sent on an official mission of investigation and trade to China. There had been limited Sino-Japanese contacts throughout those many years carried out by Chinese trading vessels that made periodic trips to Hirado and later Nagasaki, the only port open to them during most of the Tokugawa era (1600–1868); and a small number of Japanese fishermen had been shipwrecked, picked up by British or American vessels and deposited in Shanghai, though often not repatriated for many years thereafter because of the stringent travel restrictions of their homeland, for Japanese remained strictly forbidden from venturing on the seas.

Following Commodore Perry's opening of Japan, in 1853 with the resultant Treaty of Kanagawa in 1854, a few Japanese port cities gradually opened to Western trade and missionary activity. Nagasaki, the one Japanese port with a long history of merchant vessel activity, was opened to the Western powers in 1859. In short order, Western traders began plying the route between this Japanese port and the largest of China's ports, Shanghai. By this time, many Westerners had established businesses or branches of larger conglomerates in Shanghai, and the newly built Concessions where they lived had become famous for the magnificence of its buildings and grand style. News of such had been transmitted to Japan ever since the British victory in the Opium War, and the Treaty of Nanjing (1842) following it had similarly opened Shanghai, as one of five Chinese ports, to Westerners.

Japan's shogunal government, at the instigation of several high officials, decided in 1862 to send a mission abroad to investigate the conditions surrounding international trade through the microcosm of Shanghai. After all, it was much closer than travelling all the way to the countries of the West and much more condensed than the considerably more expansive continents of Europe

---

Notes for this chapter begin on page 47.

and North America. That China and Japan did not have diplomatic relations at the time would not be a problem, for the mission had no intention of visiting the Chinese capital in Beijing which was neither a thriving commercial entrepôt nor, for that matter, even open to foreigners at the time. The Japanese were only interested in trade at this juncture, as they correctly read the writing on the wall. International intercourse and commerce were the future – and Japan could join it or fall victim to it as the Chinese seemed to have done.

The mission attracted a wide variety of shogunal officials, interpreters, doctors, cooks and a large number of 'attendants', one or two for each of the samurai sent by their domains; in all, 51 Japanese made the trip. It was this last group of attendants, most of them samurai themselves and extremely well educated young men, who have left us the most detailed accounts of their ten weeks in Shanghai in the summer of 1862, and their works have been analyzed with great sophistication by a number of fine Japanese and, more recently, Chinese scholars.[1]

Because the Japanese had no recent history of open sea travel in 1862, they were faced with several options on how to make such a voyage to China materialize. Their first choice was to hire a foreign vessel and pay the foreign crew to sail them, but this proved to be far too expensive. The option of purchasing a vessel and sailing it themselves was beyond the realm of the possible at the time. The only route left open to them, then, was a kind of compromise, to buy a Western ship and then hire the crew back to navigate the waters. Thus, shogunal officials approached one Captain Henry Richardson of Great Britain who had been travelling back and forth for several years between Nagasaki and Shanghai, transporting Japanese goods to China and vice versa (and becoming wealthy as a result of it), buy his ship, the *Armistice*, and then hire him and his men back to sail it for them. The ship was renamed the *Senzaimaru* indicating, as the *North-China Herald* (1862: 1) put it at the time, 'To last a thousand years', as her name literally meant.

The one problem the new owners of the *Senzaimaru* would face in the absence of diplomatic relations with China was a legal basis for engaging in trade in Shanghai. There was simply no grounds upon which it might transpire. This difficulty was overcome with alacrity through the good offices of the Dutch trading firm of Theodorus Kroes (1822–89) which had offices both in Nagasaki and in Shanghai. The Dutch had been the sole European power allowed to trade with Japan throughout the previous two centuries and thus had a special relationship with the Japanese. Also, while there were established shogunal interpreters who accordingly knew Dutch in 1862, exceedingly few Japanese at that time knew more than a smattering of English. For a not inconsiderable fee, Kroes facilitated the commercial side of the *Senzaimaru*'s voyage, made contacts and stored goods for the Japanese and helped on the political side by offering official introductions to the Chinese bureaucracy in Shanghai.

The ten weeks the passengers on the *Senzaimaru* spent in Shanghai, from their arrival on 2 June until their return to Nagasaki in early August, were a period of brutal heat and humidity. Three crew members died, at least one from dysentery contracted as a result of inadvertently imbibing the filthy waters of the Wusong River in which everything they consumed had been washed. Many others suffered from recurrent bouts of dysentery, at a time when even the better hotels in Shanghai – such as the Astor House where they took up residence – lacked plumbing of any sort. The travel narratives penned as a result of this voyage frequently remarked on the widespread filth in the city, especially among the Chinese (e.g. Hibino 1946: 73). Lord Oliphant (1829–88) had described Shanghai three years earlier as 'the most unhealthy [port] to which our ships are sent, the sickness and mortality being greater here than even on the west coast of Africa' (Oliphant 1859: 269).

Most of the Japanese travel narratives from this voyage were the works of samurai 'attendants', none of whom were as yet out of their twenties. Several were slated for great fame in the years that followed, some for even greater posthumous fame. The one who wrote the most extensive account of his experiences and has acquired the status of the most celebrated among the 51 Japanese was Takasugi Shinsaku (1839–67), a firebrand opponent of the shogunal government's allegedly appeasing policy toward foreigners, who hailed from the domain of Chōshū. Although still a month shy of his twenty-third birthday when the mission set off, Takasugi had already at this young age acquired considerable erudition in East Asian, specifically traditional Chinese Confucian, learning; for example, he was, like his celebrated teacher Yoshida Shōin (1830–1859), an accomplished student of the literary Chinese language.[2]

Soon after boarding the *Senzaimaru*, however, he realized how underprepared he had been for the trip when he met a deck hand by the name of Godai Saisuke (Tomoatsu, 1835–1885) from Satsuma domain, like Chōshū a hotbed of anti-shogunal activity. Satsuma had been unable to get Godai's name on the passenger list in time, but he had nonetheless resourcefully managed to get himself hired in a menial capacity, despite his samurai status. He stunned Takasugi when he let the younger man in on the secret that he was studying the ship's navigational techniques and its trading routes to and from Shanghai, as well as commercial conditions in China on behalf of his home domain. Unlike Takasugi who died of illness only a few years after his return to Japan, Godai went on to become a notable figure in the Satsuma navy and, after the Meiji Restoration of 1867–68, the first president of the Osaka Chamber of Commerce.[3]

One further passenger worthy of note – space prevents a complete prosopography of the Japanese passenger list – was an 'attendant' from the domain of Hizen by the name of Nakamuda Kuranosuke (1837–1916), who would subsequently rise the post of vice-admiral in the Japanese navy.[4] Nakamuda was a great oddity for his age in that he actually knew a bit of English which enabled him to converse with the small handful of Westerners the Japanese met in Shanghai and to read a bit of the Anglophone press there – we should recall that Japan had no legal press at the time. Although Takasugi has posthumously managed to attract the lion's share of scholarly attention, Nakamuda may in fact have been the one who guided him through the streets of Shanghai and gained him entrée with the well-known British missionary, William Muirhead (1822–1900), a prolific scholar in Chinese and English and long-term resident of the city who also ran a hospital there. Muirhead was, among other things, a bibliophile, and he lent out to Takasugi and Nakamuda a number of Chinese texts he had collected concerned with current events, such as a four-volume work on the Taiping rebels, which Takasugi devoted many hours to copying out by hand (Takasugi 1974: 159–60).[5]

As fate would have it, two days after the Japanese arrived in Shanghai attacks began on the outskirts of the city by the armed forces of the Taiping leader Li Xiucheng (1823–64). Upon hearing the gunfire, Takasugi wrote in his account that he planned to try (though ultimately unsuccessful) to see something of the fighting: 'If these reports turn out to be accurate, it would make me so happy to be able to go and witness the fighting at first hand' (Takasugi 1974: 144–45, 156). There is no question that Takasugi felt an almost instinctive support for the Taipings in their struggle to topple the Qing dynasty (1644–1911), unaffected by knowledge of their devotion to Christianity, and this sympathy for their cause was only heightened when he learned that the British were aiding the Qing dynasty's efforts to crush the Taipings. His teacher Yoshida Shōin had learned of the Taiping Rebellion while in prison three years earlier awaiting execution and, unlike Takasugi, supported their suppression at that time because of his disgust for their devotion to Christianity. Takasugi, by contrast, looked beyond their alien faith to the more important fact in his eyes that they were resisting the foreigners who supported the decrepit Qing (Haga 1984: 106–09). This is in no way to conclude that he came to agree with the Taipings' core beliefs – for those he had nothing but disdain – but only that ultimately he came to blame the Qing dynasty for allowing missionaries onto their terrain in the first place, causing the spread of their noxious religion and hence the rise of the Taiping movement.

As it turned out, the principal Taiping leader Hong Xiuquan (1813–1864) ordered his troops to withdraw from Shanghai just as they were about to lay siege to the city, due to attacks by government forces on his own capital in Nanjing. Takasugi and Nakamuda would try on several occasions to locate them, but the rebels always managed to elude the two Japanese. Takasugi did observe outside the city limits Qing troops who were training to defend the city from future assaults, but he thought them a miserable sight. Just beneath the surface of all of Takasugi's observations was his fear throughout that the rampant chaos in China might spread to Japan. As he put it, 'I fear that we in Japan shall also make these grave errors', and allow the West to invade and control Japan (Takasugi 1974: 191–92).[6]

After returning home, Takasugi became an even more vigorous and stalwart opponent of accomodation with the West. Although he died on the eve of the Meiji Restoration, he became a heroic figure in death among young, disaffected Japanese as Japan sought to navigate the currents between heavy pressures from the Western powers to open up ports to commercial and missionary activity and from various domains that staunchly opposed such measures. Nowhere was this image of Takasugi truer than in his own domain of Chōshū, but he remains a household name throughout Japan, his persona reproduced in historical fiction and in historical dramas on Japanese television.[7] In later decades well after his death, at times when Western pressures on Japan were felt with particular acuteness, Takasugi was hailed as a pioneer, an important precedent for the case study described below.

Recently, there was a startling development in the ongoing interest in Takasugi and the 1862 mission to Shanghai of the *Senzaimaru* in which he played such a seminal role. In 2001 a movie version of the voyage in which Takasugi's character enjoys centre stage was discovered in the *Gosfilmfond*, the archives of the former Soviet state film industry. The movie, entitled *Noroshi wa Shanhai ni agaru* (Signal Fires over Shanghai), had certainly been known about, largely because its director was none other than Inagaki Hiroshi (1905–80), one of the most famous Japanese directors of the twentieth century, and starred the virtually legendary Bandō Tsumasaburō (1901–53, known affectionately as Bantsuma) as Takasugi himself.[8]

Noroshi wa Shanhai ni agaru was shot on location in Shanghai in 1944 and appeared late that year, a time when the city of Shanghai (along with large tracts of Chinese terrain elsewhere) was under the occupation of Japanese military forces. Although Inagaki is listed as director, the film was ostensibly a joint venture with the Chinese and was in fact co-produced by the Japanese company Dai'ei and the Chinese company Zhonghua dianying gongsi. Given the wartime conditions and the clear message of the film, however, how bilateral the production actually was remains open to serious doubt. Nonetheless, Yue Feng (1909–99), then a still young Chinese director who would subsequently rise to considerable fame in the movie business, is listed as co-director along with Hu Xinling (1914–2000), who had studied in Japan (email comm. Poshek Fu), and the film also bore the Chinese title: *Chunjiang yihen* (Lingering Resentment from Shanghai); the relationship between the Japanese and Chinese titles is vague, though it is just as unclear what the 'signal fires' of the former specifically refer to.

The film itself came out in November in China and on 28 December 1944 in Japan, less than eight months before the latter's cataclysmic defeat in World War II, and was subsequently lost. Apparently the last extant print had belonged to the Manchurian Film Company (Man'ei), created in the late 1930s under Japanese auspices on the Mainland, and was sent back to the Soviet Union following the Soviet defeat of the Japanese in Manchuria in August 1945. It thus ended up among a cache of some 55,000 films in the old *Gosfilmfond* that were discovered in 2001. The first reel of the original is still missing.

While *Noroshi wa Shanhai ni agaru* was certainly made under wartime conditions and reflected the ideology of the Greater East Asia Coprosperity Sphere – in Shanghai cinema, this specifically called for a critique of the overpowering and decadent cultural influence of the United States and Great Britain – it nonetheless was also a lot more than simple propaganda. Zhonghua dianying gongsi (literally, 'China Film Company') was in fact created in Shanghai in 1939 at the initiative of the Japanese military as a 'national policy company' (*kokusaku kaisha*), and the principal player in it was the chairman of the famed Tōwa shōji company (forerunner of the present Tōhō tōwa company), Kawakita Nagamasa (1903–1981). Kawakita was no pawn of the imperialist Japanese state. He had been raised in China, educated in Beijing, and as a result spoke Chinese effortlessly. In addition, he was profoundly drawn to the Chinese people at the time – among his film credits are movies from the 1930s with entirely Chinese casts – and at great risk to his own life, he wanted this new Chinese film company to be something more than a marionette of the Japanese military (Fu 1997: 68, 72–73; Fu 1998; Shimizu 1995; Tsuji 1987: 150–58; Yamane 2001).[9]

As a result, Kawakita saw to it that *Noroshi wa Shanhai ni agaru* would have genuine Chinese stars, directors who could hold their own on both sides, and a storyline that did not disparage the Chinese, but would somehow also cleave to the party line. Inagaki was able to secure Yahiro Fuji (b. 1904), with whom he had not worked in a decade to write the screenplay. As Yahiro later recalled, the starting point for the movie was the career of Godai Tomoatsu, but in 'researching Godai', who had left no account of his trip to Shanghai, 'I realized that I would have to examine the diaries of his fellow passenger to Shanghai, Takasugi Shinsaku …. I also had to research the biography of Nakamuda Kuranosuke …. In reading Takasugi's diaries, I discovered that he had inadvertently confronted the Taiping Rebellion' (Yahiro 1974, quoted in Takase 2000: 234).

The first reel of the film allegedly concerned the background of Takasugi's making his way onto the official passenger list of the *Senzaimaru*; as the film telescopes this history, namely his plans with fellow disaffected samurai to assassinate Nagai Uta (1819–1863), an important actor in the effort to merge the court and the military as a stopgap measure to prevent toppling the Tokugawa Shogunate. Takasugi was also active in an effort to burn down the British Legation as a protest, and this involvement led his domainal lord to whisk him out of harm's way.

The movie as we now have it begins with the *Senzaimaru* entering the port of Shanghai and navigating among countless Chinese vessels in an already thoroughly clogged harbour. Gunfire is immediately heard in the distance, and with a beaming smile on his face Takasugi explains to several of his compatriots on deck that there is a war going on. When asked what 'war' he is referring to, Takasugi replies that the Taiping Rebellion is battling the Qing dynasty (founded in 1644 by the Manchus).

We must assume that Takasugi had acquired more than a modicum of information concerning the Taipings, because his own teacher, Yoshida Shōin, had edited and translated from Chinese a recent work on that rebellion which he surely had seen. In addition, he came to know considerably more as a result of the texts provided to him by the British missionary William Muirhead (Masuda 2000: 21–22, 135–39). Soon after they reach Shanghai, there is a scene in which the Japanese taking a walk in the city witness Government troops shoot down an escaping man whose hat tumbles from his head as he falls to the ground dead – attached to the hat is a fake queue, the 'pigtail' hairstyle imposed on all Chinese males by the Manchu conquest dynasty and which all Taipings had cut off as a sign of their commitment when they joined the movement. Yahiro, Inagaki and their colleagues were thus taking a bit of artistic licence here, for no such event is reported in any of the travel narratives of the 1862 voyage, but this does effectively set the appropriate stage for the central linkage of the film – the Taipings and the radical samurai – and the licence taken here pales in comparison to what is to follow.

Throughout the film, it is Takasugi who clearly understands that the British and Americans are the real enemies of all Asians. Late one night he rescues and then befriends a fictional Taiping leader by the name of Shen Yizhou (played by the veteran, handsome actor Mei Xi, 1911–83, who was already a star of Chinese cinema at this time). Shen serves under none other than Li Xiucheng, the 'Loyal King' of the Taiping Heavenly Kingdom (played by Yan Jun, 1917–80). Shen has signed a contract with the British for weapons, and he simply does not understand Takasugi's antipathy for the Westerners. Takasugi asks him (rhetorically) if the British and Americans can be trusted, especially given the humiliating treaties they have imposed on the Chinese after the Opium and Arrow Wars. Shen claims that what is past is past, but Takasugi asks him then if he is aware of conditions in India, something about which there is virtually no way Takasugi himself could have known. Indeed, he continues, the foreign Concessions carved out of Shanghai are the first step on China's pursual of the same road to depravity. Shen is prepared to blame the alien Qing government for China's present state, and it is the goal of the Taipings, he announces, to overthrow the Manchus. Takasugi protests that he, as a Japanese, is no simple bystander to all this, because the British and Americans certainly have designs on Japan, too. Shen remains unmoved by his strangely persistent interlocutor; for, after all, both the Westerners and the Taipings are Christians. This last point only raises Takasugi's hackles, and the meeting breaks off.

Shen is actually in the city of Shanghai with a letter from Li Xiucheng to the British Consulate, and there he meets W.H. Medhurst (1822–85) who tells him that the Westerners are not going to like the Taipings' outright assault on opium dens and their ban on opium in their own liberated areas, but Medhurst thinks it the only humane approach to the deadly drug. The actor who plays Medhurst (whose name is given only as 'Orlov' in the credits) looks roughly sixty years of age. There was a senior W.H. Medhurst (1796–1857) who had written extensively on Sinitic topics, including translations of portions of the Bible into Chinese, but he was already dead by the time of this visit. Medhurst *fils* was much less accomplished and only forty years of age at the time.

When the evil British later betray the Taipings in the vilest of manners, spewing out (anachronistic) epithets among themselves for the Chinese and Japanese and literally throwing Shen out of the consulate in Shanghai, then finally, at the film's end, the message becomes resoundingly clear. Asians must stand with Asians, for the Westerners will only betray them. United they can struggle for liberation from imperialism and colonialism; divided against themselves, they will never succeed. The only Westerner painted with the least dimension to his character, Medhurst himself, tries to have his fellow British recognize the agreements they have already concluded with the Taipings, but the other British only laugh at him. He protests that the Taipings are, after all, Christians 'just like us', but again he is met with ridicule, because the Taipings, it is stridently noted to him, are not at all 'like us' – they do not have 'white skin'. When he later meets a British ship's captain who has returned to Shanghai after narrowly escaping an attack on foreign vessels in the Straits of Shimonoseki by samurai from Takasugi's home domain, the captain explains that the assault was just what the British were waiting for: an excuse to 'harm the Japs'. The camera fades with Medhurst sadly saying to himself: 'India, China, and now Japan'.

It is a bit difficult to take many of these scenes seriously now inasmuch as every single Anglophone character was portrayed by a Russian, some with accents so thick and intonation so far-fetched that it is clear they were doing little more than pronouncing words whose meanings were unknown to them. In 1944, it would have been extremely difficult to find Caucasian faces in Shanghai who were both genuinely Anglophone and willing to act in such a film; by the same token, there were many unemployed, impoverished Russophones in the city. This severe discordance to the modern (Anglophone) observer of the film would certainly not have been a major shortcoming at the time; there was no alternative, and few (if any) viewers of the time would have even been aware of the issue. The Chinese actor who plays the Taiping leader Shen Yizhou actually speaks a much better English than any of his Western interlocutors. Interestingly, though, accents aside, all of the English-language dialogue is grammatically flawless and, aside from a handful of anachronisms, indeed idiomatically genuine by nineteenth-century standards.

One of the most interesting topics for a student of Sino-Japanese interactions at this time is how the two peoples, Chinese and Japanese, actually communicated in the absence of a shared spoken language. As we learn from reading their travel narratives, they used the written literary Chinese language as a medium of discourse, thereby carrying on the ancient tradition of *bitan* (Japanese, *hitsudan*) or 'brush talks'. There is only one extended 'brush talk' in the film but it proves centrally important. Shen Yizhou's *literatus* father, Shen Changling, observes Takasugi asking a dealer in art curios about inkstones; when he realizes that Takasugi, though not a Chinese, still knows enough to seek out such culturally sophisticated artifacts, he approaches and offers to sell him just what Takasugi was looking for. A friendship ensues and they meet again at Shen's home. Shen describes with ink on paper having witnessed the death of Chen Huacheng (1776–1842), a famous figure who died fighting off the British assault 20 years earlier during the Opium War. Takasugi asks if he was killed by a British soldier and Shen replies that, in fact, it was an Indian fighting in the British army. Shen reflects on how sad this is: 'One Asian killed another'. Moments later, Takasugi says to his Japanese friend: 'Ah, Nakamuda, great chat, eh? To meet another man who wants to build a new Asia'.

Shen then asks to see the swords worn ceaselessly by the Japanese. Takasugi folds the paper on which their brush talk had transpired, puts it in his mouth (for some unexplained reason), and then unsheaths his long sword. Shen sighs admiringly: 'Japanese swords are masterpieces'. This high regard for Japanese swords is not simply propaganda, however, with the raised sword representing Japanese willingness to lead the fight of Asians against the West. There is a tradition going back at least to the eleventh century of Chinese writing poems to Japanese swords which found their way via bilateral trade onto Chinese markets (Ishihara 1960). As we have seen elsewhere, this was a virtually seamless folding of the past into the present. Also, there is evidence from the travel narratives of 1862 that a number of Chinese were curious about the ubiquitous Japanese swords.

As interesting as brush talks may be to scholars, however, they apparently did not make for great cinema. Inagaki, Yue and Yahiro thus introduced a female Chinese character named Wang Ying (played by the stunningly beautiful Li Lihua, b. 1924). Li had been trained in Peking opera and, not surprisingly, an opportunity was found for her to sing a piece from one. She later became a star in Hong Kong and Hollywood after the war. As her character explains, by virtue of a short stint in the Chinese community of Nagasaki, she can speak enough Japanese to guide the Japanese visitors as need be and serve as their interpreter. While it is conceivable that such a person existed in 1862 Shanghai – though much more likely than not it would have been a man – no such person appears in any of the travel narratives written by Japanese aboard the *Senzaimaru*. Her presence, though, serves a great facilitating role for the film; indeed, it is she who enables the 'conversation' between Shen Yizhou and Takasugi discussed above to take place.

These shortcomings are in no way sufficient, I would argue, to completely eviscerate the historical importance of *Noroshi wa Shanhai ni agaru*.[10] If there is such an element to this movie, it is the relationship between Takasugi and the Taipings, a metonym for all forward-looking Japanese and Chinese, respectively. As portrayed in this film, Takasugi forges a close bond with the Taiping leader Shen Yizhou whom he has warned against trusting the Westerners. At the film's conclusion, the Taipings are in full retreat from the city, and Shen, fleeing as well for his life, stops off at his home and says to his sister, Xiaohong (played by breathtakingly attractive Wang Danfeng, b. 1925), that he would like to see the Japanese Takasugi one more time.

She relays the message and Takasugi and his Chinese guide Wang Ying make their way to farmlands on the outskirts of Shanghai where they meet the forces of the Taiping army marching away from Shanghai after their catastrophic defeat. Somehow Takasugi finds Shen and then each delivers a soliloquy of sorts in his native tongue – which the other cannot have understood a word of – followed by the other replying (in his native tongue): 'I understand, I understand'. The thrust is clearly just the opposite, as if to say: 'I don't understand a word you have just uttered, but I now realize that what is truly important is that Asians band together against the Westerner invader'. One further historical problem with any such exchange even being conceivable is that Li Xiucheng and his forces were not in full retreat from Shanghai until several weeks after Takasugi and his colleagues set sail home for Japan in the first week of August (Jen 1973: 448–49, 458–59); it is accurate, though, that promises proffered earlier by French and British officials were reneged upon. The United States played little part in all this, although it comes in for considerable vitriol in the film, obviously due to the exigencies of the times in which the film was made, as opposed to that of the Taiping Rebellion itself.

The encounter and exchange between Takasugi and Shen are, of course, pure fantasy. As we have seen, Takasugi never met any Taipings; if he had, they surely would have made more than a cameo appearance in his narrative. Any Taiping officer walking about the streets of Shanghai, as Shen Yizhou does in full, bizarre Taiping regalia, would have been summarily executed. Nonetheless, Takasugi's own biography and his account of his time in Shanghai fit with eerie precision into the framework of 1944 East Asian power politics. Anachronisms aside – and there are any number of them in the film, such as the Hinomaru, the contemporary Japanese flag, flying alone atop the *Senzaimaru* as the 'national' Japanese flag as it comes into Shanghai harbour – he warned against any accomodation with Westerners, despised their presence in Shanghai and their haughtiness with respect to the Chinese, and worried more than anything else that such a fate as had befallen China might come to Japan. Although he did not use the word 'colony' – such a term did not exist *per se* in the Japanese lexicon of 1862 – he did refer to Shanghai as a 'dependency of Great Britain' ('*Dai-Ei zokkoku*'), and the colonial implications are eminently clear (Furukawa 1973: 80; Ikeda 1966: 119; Naramoto 1965: 106–15; Takasugi 1974: 178, 185; Tanaka 1991: 244).

Thus, there is considerable evidence to see Takasugi as an early anti-imperialist icon, and perhaps he would have supported a Japanese-led, pan-Asian movement, had such a confederation ever come together in his lifetime. There is, however, a huge leap from there to Takasugi as (figuratively or, in this instance, literally) supporting the Taipings in their quest to overthrow the Qing dynasty. In his 1983 memoirs, this is how director Inagaki recalled the background and storyline of the film:

> Bantsuma and I made the film *Noroshi wa Shanhai ni agaru*, a joint Sino-Japanese venture, in Shanghai, and in it Takasugi Shinsaku appears as protagonist. With the Qing (China) defeated in the Opium War and forced to sign humiliating treaties with Great Britain, China had lost its subjectivity. Under these circumstances, revolutionaries known as the 'Taipings' rose up with their Christian ideology and attacked the Qing government. Takasugi Shinsaku, Godai Tomoatsu, Nakamuda Kuranosuke, and their associates travel to Shanghai to purchase a Western-style ship, and not indifferent to what appears to be a civil war, they foresee that Japan might as well be invaded by men from Great Britain and the United States. This story was not to be a sabre-rattling swashbuckler starring Bantsuma, nor was I going to shoulder the burdens of the Pacific War ...

Joshua A. Fogel

> Bantsuma and I made this film over an eight-month period in Shanghai. When it was first released, both Japan and Shanghai were in the midst of air raids, and people had no time to go watch movies. But, at least for those of us who made it, we were calmly watching the rise and conclusion of the Pacific War, just as Takasugi in the drama was objectively observing the Taiping Rebellion in the Qing period. (Inagaki 1983: 281–82)[11]

There is one scene of comic relief in this film which actually finds the Japanese the butt of a comic mishap. One morning Godai Tomoatsu (portrayed by Tsukigata Ryūnosuke, 1902–70) is trying to get two overly helpful Chinese bellboys at the hotel in Shanghai to make tea for himself and Nakamuda Kuranosuke (played by Ishiguro Tatsuya, 1911–65).[12] He finds himself unable to convey this simplest of words to the Chinese: *ocha* in Japanese, *cha* in Chinese, but with different intonation. Takasugi comes downstairs, intrudes on the confusion and takes over, repeating the word *ocha* a number of times. When the Chinese start mimicking the Japanese word, he replies several times: '*Kore da*' ('That's it'). Then, the Chinese start repeating the words '*kore da*', until they finally believe they have discovered what these strange visitors want to drink: *gaoliang jiu*, a particular potent alcoholic beverage made from sorghum. Somewhat later, Godai is seen taking a sip and spitting it up all over himself. The two Chinese bellboys were played by Han Lan'gen (1909–1982) and Yin Xiucen who had become famous as the 'Oriental Laurel and Hardy', respectively, in a number of movies from the 1930s and 1940s (Fang 2001).

Whatever one may think of the cinematic qualities of *Noroshi wa Shanhai ni agaru*, participation in this production cannot have stood the Chinese actors and staff in good stead after the conclusion of the war and the founding of the People's Republic of China. As noted, Yue Feng, the Chinese director, went on to a career for some years in Hong Kong where he made a number of movies starring several Chinese actors from the 1944 joint venture under study here. His colleague Hu Xinling worked in the Hong Kong and Taiwan film industry. The married couple Li Lihua and Yan Jun both appeared in Yue Feng's 1952 film *Xin Hongloumeng* (The New Dream of the Red Chamber); Yan appeared in at least four other movies directed by Yue in the 1949–50 period alone. Wang Danfeng worked in Hong Kong for several years before returning to China in the early 1950s, and there she was able to enjoy a lengthy career in cinema, though she was the exception that proved the rule.[13]

Mei Xi's filmography is difficult to ascertain in full, but with a number of lengthy lacunae, he managed to act for most of his life in mainland China. Han Lan'gen, too, stayed on in the People's Republic after 1949 and suffered through a number of Communist campaigns, never able to get his derailed career back on track. Of the Russian actors who played all the Anglo-American and French roles, nothing is immediately forthcoming except several of their surnames (Orlov, Moskarenko, Serebanov). It would not be terribly surprising if they all found work in other professions after the war. Their performances in this production are, from a thespian point of view, execrable – to say nothing of their English.

The voyage of the *Senzaimaru* to Shanghai in 1862 marks the dawn of modern Sino-Japanese relations in government, commerce and culture. It would have marked a historical milestone even if it had not had an impact back in Japan, when Takasugi found the embers of his anti-foreign hostility flaring up again. This led, in part, to the further awakening of radicalism in Chōshū domain in what eventually brought on the collapse of the Tokugawa regime a mere five years after dispatching this mission and the rise of a unified Meiji government. It is this latter part of the story that had legs, that continued to exercise an impact on angry young Japanese over the decades that followed, and that ironically provided just the right mix for the state-sponsored film of 1944, *Noroshi wa Shanhai ni agaru*. Indeed, Yahiro, Inagaki and their production colleagues need not have tinkered with the history of the Taiping Rebellion and its links to the Japanese to have told the story they ultimately wished to tell. The truth would have sufficed, though it might not have made as good a story on film. Binding these two seminal events – radical anti-shogunal samurai and the Taiping rebels – gave the film an added punch, as described above.

Viewing the film now, nearly 60 years after its initial release, one is struck by a certain bizarre quality in its message. Putting aside the absurdities of the Russian actors, the anachronisms in speech and practice, and the datedness of the acting styles of the Chinese and Japanese performers, the message of the need for Chinese and Japanese to come together to ward off Western advances against them both in the middle of the nineteenth century retains considerable poignancy. This in no way is meant to validate Japanese imperialism; that is a given in 1944. Nonetheless, just as Takasugi 'spoke' to Inagaki and his collaborators over 80 years after he set sail for Shanghai, so he continues to speak to us nearly 60 more years after the film first appeared.

Joshua Fogel is professor of history at York University in Toronto, Canada. He specializes in the cultural dimension of Sino-Japanese relations in the 19th and 20th centuries. He has written, edited, and/or translated sixty-five books. His most recent monographs include: *Maiden Voyage: The* Senzaimaru *and the Creation of Modern Sino-Japanese Relations* (University of California Press, 2014); *Japanese for Sinologists: A Reading Primer with Glossaries and Translations* (University of California Press, 2017); and *Friend in Deed: Lu Xun, Uchiyama Kanzō, and the Intellectual World of Shanghai on the Eve of War* (Association for Asian Studies, 2019). In addition he is the founding and continuing editor-in-chief of *Sino-Japanese Studies*.

Joshua A. Fogel

Notes

1. The best and most comprehensive recent work in Japanese is Miyanaga (1995), and in Chinese Feng (2001). My own work on the voyage of the *Senzaimaru* can be found in Fogel (1994) and (1996: 43–65 ('First Contacts: The Travelers Aboard the 'Senzaimaru' and Other Early Accounts')).
2. Takasugi's accounts are known under the collective rubric of *Yū-Shin goroku* (Five Accounts of a Voyage to China); they have been published several times, most definitively in an edition edited by Tanaka Akira (1991: 209–86); an earlier edition is Takasugi Shinsaku 1974: 141–216.
3. Godai left no narrative of his experiences in Shanghai (see Miyamoto 1981; Okita 1942; and Tanaka 1921).
4. Nakamuda did leave an account, entitled *Shanhai tokō kiji* (Diary of a Crossing to Shanghai), although it is extremely difficult to find. It was excerpted and used extensively by his biographer, Nakamura Kōya (Nakamura 1919; see also Haruna 1997).
5. On Nakamuda's relative importance, see Haruna Akira (1997).
6. This point has been commented on by a number of scholars, among them: Etō (1970); Satō (1984); and Fogel (1996: 53–54).
7. Takasugi's life is the subject of a four-volume novel by the greatest of modern Japanese historical novelists, Shiba Ryōtarō (1923–1996) (Shiba 1971). Each year Japanese educational television (NHK) shows in 52 one-hour weekly segments a work of historical fiction; in 1977 they serialized another of Shiba's novels, *Kashin* (God of Flowers) (Shiba 1972), which had no signficant mention of Takasugi, though in the television drama Takasugi played an important featured role with material drawn from *Yo ni sumu hibi*.
8. Inagaki is probably most famous, at least in the West, for the *Samurai Trilogy* (1956), the five-hour, 22-minute marathon 'biopic' concerning the figure of Miyamoto Musashi (1584–1645) and starring Mifune Toshirō (1920–1997) in the leading role. Inagaki and Mifune would work together in a number of important films. Bandō – born Tamura Denkichi and widely known as Bantsuma – began his own production company in 1925 under the latter name and starred in numerous period pieces. The rediscovery of this film coincided with the centenary of his birth, and there were a number of film series and photographic displays in Japan to honour him in which this film was featured (see Segawa 1977).
9. Poshek Fu (Fu 1997: 68–69) suggests that Kawakita was more self-concerned than the preceding may indicate, that he wanted to produce entertainment and was concerned about alienating the bulk of the Shanghai film world who might flee to the hinterland should the Japanese apply too restrictive a policy on film-makers. He thus pushed for the creation of a single, centralized film industry in Shanghai under his Zhonghua dianying gongsi which would be of, by and for the Chinese people (see also Fu 1994). See Fu (1997) further for details of Kawakita's relationship to Zhang Shangkun, the man in charge of the day-to-day operations of Zhonghua dianying gongsi.
10. Poshek Fu (1997: 79) is much more critical of the film as a propagandistic effort to rouse anti-Western sentiment. Because the film was only recently rediscovered, Fu was unable actually to view it and was forced to rely on interviews with surviving Chinese actors, such as Lü Yukun, who was still smarting four decades later.
11. Inagaki tried to entice the extraordinarily popular actress Yamagichi Yoshiko (b. 1920) to star in the female Chinese lead in the movie. At the time, she was known by the Chinese name Li Xianglan (J. Ri Kōran) and spoke Chinese fluently; although bearing Japanese citizenship, she actually spoke Japanese with a Chinese accent by virtue of having been born and raised on the mainland – and after the war, many thought she had been a Chinese who had agreed to make pro-Japanese films. She was only too happy at the time to take part in a Sino-Japanese joint venture, but the collaboration never materialized. See Inagaki 1983: 203–04.

12. All three principal Japanese leads – Bandō, Tsukigata and Ishiguro – were well-known period-piece actors at the time of filming. Only Ishiguro was remotely close in age to the character portrayed. Bandō and Tsukigata were actually twice the age of their characters and it shows, particularly with the former.
13. The information in this paragraph was drawn from a number of Chinese and Japanese websites devoted to film history, such as that of the 'Hongse jingdian' ('Red classics'), which features a biographical portrait of Wang Danfeng that makes no mention whatsoever of *Noroshi wa Shanhai ni agaru*, (http://202.108.249.200/ specials/hsjd/sanji/wangdanfeng.html) and that of Chinanews.com which discusses Wang's career (http://www.chinanews.com.cn/zhuanzhu/2001–10–09/ 648.htm), as well a number of Japanese sites marking retrospectives of Bandō Tsumasaburō's films (all accessed in early 2004).

References

Etō Shinkichi (1970) 'Nihonjin no Chūgokukan: Takasugi Shinsaku ra no baai' (Japanese views of China: The case of Takasugi Shinsaku and others), in Fukushima Masao (ed.) *Niida Noboru hakase tsuitō ronbunshū, daisankan: Nihon hō to Ajia* (Essays in memory of Professor Niida Noboru, vol. 3: Japanese law and Asia), pp. 53–71. Tokyo: Keisō shobō.
Fang Baoluo (2001) 'Han Langen', in http://www.gstage.com/cgi-bin/f_article.cgi?article = 2721.
Feng Tianyu (2001) *'Qiansuiwan' Shanghai xing: Ribenren 1862 nian de Zhongguo guancha* (The *Senzaimaru*'s trip to Shanghai: Japanese views of China in 1862). Beijing: Shangwu yinshuguan.
Fogel, J.A. (1994) 'The Voyage of the *Senzaimaru* to Shanghai: Early Sino-Japanese Contacts in the Modern Era', in J.A. Fogel, *The Cultural Dimension of Sino-Japanese Relations: Essays on the Nineteenth and Twentieth Centuries*, pp. 79–94. Armonk: M. E. Sharpe.
Fogel, J.A. (1996) *The Literature of Travel in the Japanese Rediscovery of China, 1862–1945*. Stanford: Stanford University Press.
Fu, P. (1994) 'Struggle to Entertain: The Political Ambivalence of the Shanghai Film Industry under Japanese Occupation, 1941–1945', in Law Kar (ed.) *Cinema of Two Cities: Hong Kong-Shanghai*, pp. 50–62. Hong Kong: Urban Council.
Fu, P. (1997) 'The Ambiguity of Entertainment: Chinese Cinema in Japanese Occupied Shanghai, 1941 to 45', *Cinema Journal* 37, 1 (Fall): 66–84.
Fu, P. (1998) 'Projecting Ambivalence: Chinese Cinema in Semi-Occupied Shanghai, 1937–41', in Wen-hsin Yeh (ed.) *Wartime Shanghai*, pp. 86–109. London and New York: Routledge.
Furukawa Kaoru (1973) *Takasugi Shinsaku*. Tokyo: Shin jinbutsu ōraisha.
Haga Noboru (1984) 'Ahen sensō, Taihei tengoku, Nihon' (The Opium War, the Taiping Rebellion, Japan), in *Chūgoku kin-gendai shi no sho mondai: Tanaka Masayoshi sensei taikan kinen ronshū* (Problems in modern and contemporary Chinese history, essays commemorating the retirement of Professor Tanaka Masayoshi), pp. 87–123. Tokyo: Kokusho kankōkai.
Haruna Akira (1997) 'Nakamuda Kuranosuke no Shanhai taiken: *Bunkyū ninen Shanhai kō nikki* o chūshin ni' (Nakamuda Kuranosuke's experiences in Shanghai: On the 'Diary of a Trip to Shanghai in 1862'), *Kokugakuin daigaku kiyō* 35 (March): 57–96.
Hibino Teruhiro (1946) *Zeiyūroku* (A record of warts and lumps), in *Bunkyū ninen Shanhai nikki* (Diaries of Shanghai in 1862). Osaka: Zenkoku shobō.
Ikeda Satoshi (1966) *Takasugi Shinsaku to Kusaka Genzui, henkakki no shōnenzō* (Images of youth in an era of transformation: Takasugi Shinsaku and Kusaka Genzui). Tokyo: Daiwa shobō.
Inagaki Hiroshi (1983) *Nihon eiga no wakaki hibi* (The early days of Japanese cinema). Tokyo: Chūō kōronsha.

Ishihara Michihiro (1960) 'Nihon tō shichishu: Chūgoku ni okeru Nihonkan no ichimen' (Seven poems on Japanese swords: One Chinese view of Japan), *Ibaraki daigaku bunrigakubu kiyō, jinbun kagaku* 11 (December): 17–26.
Jen Yu-wen (1973) *The Taiping Revolutionary Movement*. New Haven: Yale University Press.
Masuda Wataru (2000) *Japan and China: Mutual Representations in the Modern Era*, trans. J.A. Fogel. Richmond: Curzon Press.
Miyamoto Mataji (1981) *Godai Tomoatsu den* (Biography of Godai Tomoatsu). Tokyo: Yūhikaku.
Miyanaga Takashi (1995) *Takasugi Shinsaku no Shanhai repotto* (Takasugi Shinsaku's report on Shanghai). Tokyo: Shin jinbutsu ōraisha.
Nakamura Kōya (1919) *Nakamuda Kuranosuke den* (Biography of Nakamuda Kuranosuke). Tokyo: Nakamuda Takenobu.
Naramoto Tatsuya (1965) *Takasugi Shinsaku, ishin zenya no gunzō* (Takasugi Shinsaku, a portrait on the eve of the Meiji Restoration). Tokyo: Chūō shinsho.
The North-China Herald (1862) 619, 7 June.
Okita Hajime (1942) 'Godai Tomoatsu to Shanhai' (Godai Tomoatsu and Shanghai), *Shanhai* 1 (January): 16–19.
Oliphant, L. (1859) *Narrative of the Early of Elgin's Mission to China and Japan, 1857–1859*, vol. 1. Edinburgh and London: William Blackwood and Sons.
Satō Saburō (1984) 'Bunkyū ninen ni okeru bakufu bōekisen Senzaimaru no Shanhai haken ni tsuite' (On the sending to Shanghai in 1862 of the shogunal trading vessel, the *Senzaimaru*), in Satō Saburō, *Kindai Nit-Chū kōshō shi no kenkyū* (Studies in the history of modern Sino-Japanese relations), pp. 67–96. Tokyo: Yoshikawa kōbunkan.
Segawa Ken'ichirō (1977) *Bandō Tsumasaburō*. Tokyo: Mainichi shinbunsha.
Shiba Ryōtarō (1971) *Yo ni sumu hibi* (The world in which we live). Tokyo: Bungei shunjū.
Shiba Ryōtarō (1972) *Kashin* (God of flowers). Tokyo: Shinchōsha.
Shimizu Akira (1995) *Shanhai sokai eiga watakushi shi* (A personal history of films in the Shanghai Concessions). Tokyo: Shinchōsha.
Takase Masahiro (2000) *Wagagokoro no Inagaki Hiroshi* (Inagaki Hiroshi in our minds). Tokyo: Waizu shuppan.
Takasugi Shinsaku (1974) *Yū-Shin goroku* (Five accounts of a voyage to China), in Hori Tetsusaburō (ed.) *Takasugi Shinsaku zenshū* (Collected works of Takasugi Shinsaku), vol. 2, pp. 141–216. Tokyo: Shin jinbutsu ōraisha.
Tanaka Akira (ed.) (1991) *Kaikoku* (Opening the country). Tokyo: Iwanami shoten.
Tanaka Toyojirō (1921) *Kindai no ijin: Ko Godai Tomoatsu den* (A great modern man: Biography of the late Godai Tomoatsu). Ōsaka: Tomoatsukai.
Tsuji Hisakazu (1987) *Chūka den'ei shiwa, ichi heisotsu no Nit-Chū eiga kaisōki* (Tales from the history of Chinese cinema, one soldier's memoirs of Sino-Japanese films). Tokyo: Gaifūsha.
Yahiro Fuji (1974) *Jidai eiga to gojūnen* (Period movies and fifty years). Tokyo: Gakugei shorin.
Yamane Sadao (2001) 'Yamane Sadao no otanoshimi zeminaaru' (Yamane Sadao's marvellous seminar). Liner notes to the video of *Noroshi wa Shanhai ni agaru*, distrib.: Kinema Club.

Part II
Visualizing Otherness

Chapter 4

SEEING A DIFFERENCE
Spectacles of Otherness in Eighteenth-Century Illustrated Travel Books

Julia Thomas

The scene is set. Father Time reaches up to tie back the heavy swathes of curtain and reveals the extraordinary spectacle. But we are not the only viewers. Looking intently at what lies before her is an angelic scribe, who, quill in hand, records what she sees. To us, the angel's words are illegible, the marks of ink only just visible along the top edge of the large manuscript she holds. Her divine text, however, has an earthly counterpart: the very Volume that we have open before us.

This image forms the frontispiece of a book that sets out to describe the weird and wonderful sights that the angel witnesses: Cornelius Le Bruyn's *Travels into Muscovy, Persia, and Part of the East-Indies* (Illustration 4.1). Originally published in Amsterdam in 1714, Le Bruyn's account ran through several editions during the following decades and was translated into English and French. The eighteenth century, after all, was the age not just of the intrepid explorer, but also the armchair traveller. While Captain Cook sailed the islands of the South Pacific and young men took the Grand Tour of Europe, the less adventurous need venture no further than the nearest bookshop or circulating library.

In this period the appeal of travel books itself crossed geographical divides. Whether they recounted the scientific explorations undertaken by the likes of Cook, or private, even imaginary, voyages, these texts were read in different countries and continents and constituted one of the most popular literary genres. Part of the reason for Le Bruyn's particular success, though, came from the fact that his work was lavishly illustrated, containing over 300 copper plates. The number and quality of images in this and other published travel accounts was used to market these often folio-sized works, to make them into valuable and valued items, although the desire for pictures during the century was such that smaller and cheaper travel books also began to be illustrated. Le Bruyn's frontispiece, designed by Bernard Picart, the influential artist who would go on to

Julia Thomas

Illustration 4.1 Bernard Picart, frontispiece for Cornelius Le Bruyn, *Travels into Muscovy, Persia, and Part of the East-Indies*, vol. 1, London, 1737.

picture the ceremonies and religious customs of the world in seven Volumes (1723–1738), whets the appetite for the beautiful plates that follow.

Picart's engraving does more than merely advertise the content of the book, however. Despite its iconographic gesture towards the tradition of the *speculum mundi*, it stands almost as a paradigm of eighteenth-century illustrated travel accounts with all their visual delights and troubling inconsistencies. The architectural ruins, native inhabitants, wildlife and ship-at-sea that adorn this plate are precisely those images that jostle for attention on the pages of such texts. There is a sense of urgency in this assemblage of pictorial details. It is no coincidence that Father Time, in his struggle with the theatrical curtains, has dropped his scythe to the floor and is standing on an overturned hourglass. Time has stood still, and just long enough for the angel to record her vision. A moment later and all will have disappeared from her sight, or changed beyond recognition.

It is this awareness of transience that gives the illustrated travel book its pathos and appeal. The representation of the woman dressed in the native garb of Russia on the left of the plate takes on a particular resonance when set alongside Le Bruyn's later comments on another picture of the same famale: 'Time has wrought great changes in this Empire, and especially since the Czar's return from his travels. He immediately altered the fashion of dress ... As this great alteration may in time blot out the remembrance of the ancient dress of the country, I painted the dress of the Ladies upon canvass' (Le Bruyn 1737: 46).

Pictures, then, have a duty to record aspects of a changing environment, but this does not mean that they offer a unified perspective. Indeed, the Russian woman represented in the frontispiece amalgamates divergent traditions: her fur-trimmed robe, the representation of her 'real' outfit, is strangely at odds with the bare feet she sports in deference to the classical ideal. What I want to argue here is that the eighteenth-century illustrated travel book is characterised by this multiplicity and diversity. Not only does it contain images of foreign people and places, but it works to expose the very operations of difference by constructing and subverting the binary categories of fact and fiction, word and image, self and other.

Fact and Fiction

Picart's frontispiece is not a place for hiding. The curtain draped down the side of the plate tantalisingly refers to what might lie behind it, but we know that even if Father Time were to gather its folds outside the margins of the frame, there would be nothing else to see. The 'mirror' of the world that this image represents is one that seems to reveal all. To the patient observer, the truth of people, places, customs and costumes will be laid bare. Knowledge can be acquired. Facts can be attained. And all by looking and recording. No doubt it was this search for the truth that led to the travel narrative's success. This is suggested by the Swedish naturalist, Anders Sparrman, in his preface to *A Voyage to the Cape of Good Hope*, where he relates the popularity of the genre to the eighteenth-century 'disposition to enquire into facts' (Sparrman 1785: iii).

On the pages of the illustrated travel book, however, this enquiry 'into facts' frequently becomes an enquiry into the very possibility of facts, of distinguishing between the 'truthful' and the imaginative. In these texts and pictures the notion of authenticity is up for debate. And even Le Bruyn takes a part: 'I have made it an indispensable Law to my Self', he asserts, 'not to deviate in any respect from the Truth' (Le Bruyn 1737: preface). But this proclamation, repeated in almost every travel book of the period, works not only to privilege the idea of the truth, but also to expose it to the slippages and instability of language. What does it signify, then, and what are the implications for the meanings of truth, when fictional accounts of voyages describe themselves as factual?

So conventional was the affirmation of legitimacy in travel books that their parodies and satires exploited the same literary trope. A letter from the publisher to the readers of *Gulliver's Travels* remarks that 'There is an Air of Truth apparent through the whole; and indeed the Author was so distinguished for his Veracity, that it became a sort of Proverb among his Neighbours at *Redriff*, when any one affirmed a Thing, to say, it was as true as if Mr. *Gulliver* had spoke it' (Swift 1727: 31). Another famous imaginative travel account, published in 1785, went on to use this quotation in its own preface, adding that 'The Editor of these Adventures humbly hopes they will also be received with the same marks of respect, and the exclamation of THAT'S A MUNCHAUSEN! given hereafter to every article of authentic intelligence' (Raspe 1793: x).

In the case of *Gulliver's Travels* and the adventures of Baron Munchausen, of course, the allusion to the truth of the texts is highly ironic and adds to their comic potential. It was easy to recognise the fictional, to spot the difference between descriptions of a race of giants and Australian natives. Or was it? Gulliver's account of the Brobdingnags bears some resemblance to the Patagonian giants first seen by John Byron on his travels in the Pacific. By the time John Hawkesworth compiled his Volumes of explorations in 1773, he concluded that the testimony of navigators 'of unquestionable veracity', who had measured the Patagonians, would put an end to any doubts about their genuineness (Hawkesworth 1773: xvi). But, despite Hawkesworth's optimism, the actual existence of the Patagonian giants remained indeterminate. They were both imaginary and real: the stuff of myth and legend, and the objects of 'discovery' by trustworthy explorers. What the Patagonians reveal is the fragility of the distinction between truth and falsehood, the fact that 'fact' cannot easily be fixed or defined, least of all by its apparent difference from fiction. These oppositions always contain the trace of each other.

But do pictures provide the missing proof, evidence that people, places or objects were really there before the artist's sketchbook? Perhaps Hawkesworth believed they did, for he accompanies his narrative of the Patagonians with a fold-out image of Byron meeting them. The veracity of the pictures in travel books was no less emphasised than that of the words, especially if they were drawn from life or 'on the spot'. Barbara Maria Stafford has recently argued that such images demonstrate the 'fashioning of a "scientific gaze"', an attempt to represent the material world as accurately as possible and without the lure of the imagination (Stafford 1984: 52). This certainly seems to be the case in books such as *Travels in Upper and Lower Egypt*, in which the 40 illustrations are described as 'exact' (Sonnini 1799: xi), as well as Le Bruyn's own drawings, which he boasts are 'conformable to those Originals which are still to be found on the Spot' (Le Bruyn 1737: preface). But the ways in which these images achieved their apparent accuracy were themselves disparate. So much so that when William Gilpin published his *Observations on the River Wye* in 1782 he commented that the scenes were 'hastily sketched' rather than minutely rendered in order that they would appear 'just as they struck the eye at first' (Gilpin 1800: viii. 1800 edition of orginal 1792 text used here).

Yet Gilpin himself was very aware that pictures were not mere reflections of what they depicted. The images that appeared in travel books were sometimes 'touched up' in order to make them suitable for publication. This was a necessity in the case of rough drawings, which were often done quickly at the scene. Many of the images brought back from Cook's journeys were subject to such revision, being sent to artists in London, who made finished paintings of the sketches. The gap between the original image and the published version was further widened by the technique used for duplication. Alexander Dalrymple, who was rejected in favour of Cook as leader of the South Seas expedition, complained that the drawings of Alexander Buchan, the landscape and figure artist on the *Endeavour*, were ruined in their reproduction: '*Buchan's drawings* of the people of Terra del Fuego convey a perfect idea of their costume; your [Hawkesworth's] *engraving* shews what they *may be* when a French dancing-master has *taught* them *attitudes*' (Dalrymple 1773: 26).

In a period that experimented with the mechanical reproduction of images, the 'original' impression was sometimes lost sight of. As Walter Benjamin has argued, 'even the most perfect reproduction of a work of art is lacking in one element: its presence in time and space, its unique existence at the place where it happens to be' (Benjamin 1968: 214). This 'lack' is exacerbated in the case of the illustrated travel book, which is characterised specifically in terms of its 'place', its representation of a specific 'time and space'. Indeed, in the eighteenth century the means by which a picture was copied could dictate the subject matter and how it was represented. Gilpin's drawings were published as aquatints, an intaglio process invented by Jean Baptiste Le Prince around 1768. The aquatint, produced by repeatedly immersing a resined metal plate in acid, which bites out the areas to hold the ink, was able to create the tonal effects usually associated with a fine watercolour. However, while Gilpin agreed that the process came nearer than any other to the softness of drawing with a pencil, he was concerned that it took control over the appearance of the image out of the hands of the artist and was not always able to ensure a just gradation of light and shade.

It was not only the mode of reproduction that generated the meanings of illustrated travel books. There were other influences that determined the choice of material and the way it was depicted. For Gilpin, this was the idea of the 'picturesque'. Coming to prominence in the late eighteenth century (Gilpin, in fact, is credited for introducing the term into British aesthetics), picturesque images sought to convey the type of natural beauty that was suitable for 'picturing', inspiring a sensitive appreciation in their observers. Gilpin's scenes show visually striking, contoured landscapes, accentuated by subtle shading and lighting contrasts. His idealised views also draw on the landscape paintings of Claude Lorrain, whose work had become well known in Britain in the 1770s with the publication of prints of his pictures.

Even a 'hasty' sketch drawn on the spot, then, is never entirely free from the prior artistic conventions and assumptions that undermine the image's claim to absolute truth or immediacy. Gilpin's very idea that his pictures should appear 'just as they struck the eye at first' is informed by the 'pure' and direct contemplation of nature advocated by the picturesque. These influences, of course, do not have to be conscious, but are a result of the artist's and image's location in a culture that is saturated with representations. Dalrymple, it seems, had a point when he identified the staged 'attitudes' of some of the figures in the plates for Cook's voyages. But this might have had as much to do with the artist as the engraver. William Hodges, the draughtsman who sailed on Cook's second voyage, positions his natives in classical poses and landscapes (Illustration 4.2), a consequence, perhaps, of his apprenticeship with the painter Richard Wilson.

These cultural interactions, of course, were inevitable. As Rüdiger Joppien and Bernard Smith have remarked, it was only natural that European travellers should draw on their Christian and classical heritage and interpret 'the unknown in terms of the known' (Joppien and Smith 1985: 8). What these influences also suggest is the impossibility of creating a mirror image of the world. A picture in a book is just that: a picture, a representation, not the real thing it depicts. The sheer number of illustrated travel narratives in the eighteenth century, from the explorative accounts of Cook, Anson and Bougainville to

Seeing a Difference

Illustration 4.2 William Hodges, *Family in Dusky Bay, New Zealand*, from James Cook, *A Voyage towards the South Pole, and round the World*, vol. 1, London, 1777.

59

'polite' European travels, and the diversity of the images they contained, meant that no book was ever entirely original, but made up of conventions adopted by other travel books. Different botanical plates might show diverse flora and fauna, but stylistically they were alike, with minimal background, bold, sharp lines and close attention to detail. Similarly, when artists turned their attention to the ruins of an ancient city, the likelihood was that they would represent it in the intricate manner and geometric perspective popularised by Piranesi or the Scottish architect, Robert Adam.

This cultural circulation of images works to erode the distinction between fictional and factual travel accounts, resulting in what Jean Baudrillard calls a 'simulacrum', a copy that refers only to other copies and undermines the very idea of an original (Baudrillard 1994, *passim*). In order to draw an imaginary native from the moon, the artist of an illustrated travel book would look, even if unconsciously, at representations of natives closer to home. This, in fact, was precisely the pictorial strategy adopted in an illustration of an alien that appeared in Baron Munchausen's *Gulliver Revived*, which despite its scaly skin and head held under its arm, is drawn in the manner of an ethnographic portrait. Dominating the picture space, with little detraction in the form of background detail, this strange figure stands as the representative of its species and is shown in close-up for our examination. In this sense, the inhabitants of outer space are not so different from the costumed Russian women drawn by Le Bruyn.

Word and Image

But perhaps it is difference itself that is questioned in the eighteenth-century illustrated travel book. The 'otherness' of the inhabitants of outerspace depends on a visual reference to and comparison with contemporary pictures of natives. Difference here is not natural or essential, but is constituted in the images themselves. The images, however, do not work alone. The engravings that appear in *Gulliver Revived* come with captions. These textual 'titles' can work to guide a 'reading' of the scene: without the words, we might not recognise what the pictures represent; or they can undermine the possibility of any fixed interpretation: words are subject to the same slippages of meaning as images.

It is this dialogue between the textual and visual that defines the illustrated travel book. The very fact that this hybrid genre was so popular and successful suggests that words and pictures complemented each other and were regarded as equally important in the creation of travel accounts. Certainly, there was an attempt to secure the relation between the two modes of representation. On a physical level, this was the responsibility of the bookbinder, the instructions to the binders that these texts often contain suggesting an anxiety to position the image as close as possible to its textual analogue. And if the binders failed, there was always the cross-referencing in the text itself. Some authors, like Le Bruyn, refer directly to the plates in the narrative; others incorporate words within the space of the image, whether in the form of captions, alphabetical keys (adopted in architectural drawings) or descriptions of the features. A picture of a pass through the Lueg mountain in Salzburg, published in Johann Georg Keysler's *Travels through Germany*, contains inscriptions on the landscape itself: 'The Clausen', 'Road from Salzburg and Golling', and 'River Salza' are all labelled within the picture (Keysler 1760: facing p. 47). It is no wonder, therefore, that travel writers compared their activity to that of artists: 'for as a painter, to be a master in his art, ought to know the propriety and force of all sorts of colours,' asserts Louis Daniel Le Comte in his illustrated account of China, 'so whoever undertakes a description of the people, arts and sciences, and religions of the new world, must have a large stock of knowledge' (Le Comte 1739: preface).

Le Comte's correlation between travel writing and painting might have seemed familiar to eighteenth-century readers as part of an aesthetic tradition that identified the shared characteristics of the 'sister' arts. '*Ut pictura poesis*' ('as is painting, so is poetry'), taken from Horace's *Ars Poetica*, was the slogan of this method of associating the arts, which had been influential for centuries and extended beyond the generic categories of poetry and painting. Word and image were intertwined: they had same artistic aims and objectives and could represent the world in similar ways. Or so it seemed. The reaction against *ut pictura poesis* came in 1766 in the form of *Laocoon*, a book written by the German critic and dramatist, Gotthold Lessing. Lessing stressed difference rather than sameness, the fact that poetry and painting made their meanings in distinct ways: whereas poetry is temporal and should represent movement in time, painting is spatial and should select and represent just one moment (Lessing 1766).

The illustrated travel narrative is caught in the middle of these theoretical conflicts and often comes to embody them. These pictures were, in a sense, always breaking the rules: they were fixed in space, but were designed to show movement; they signified in their relation to the text, but could also work against it. Some plates seem almost to revel in these inconsistencies, to celebrate the madness in their method. The images in Edward Cavendish Drake's collection of voyages owe their greatest debt to imaginative travel books and take story-telling to new heights, in defiance of Lessing's claim that pictures should avoid showing temporal events. Because these images were not drawn on the spot but especially commissioned for this edition, which was published centuries after some of the voyages had taken place, they had a certain freedom in how they represented the scenes. The focus here is on situations

Julia Thomas

rather than landscapes, on narratives rather than factual information, as demonstrated in the picture that accompanies the following account of Captain Woodes Rogers' landing in California (Illustration 4.3):

Illustration 4.3 *The Landing of Captain Rogers's men at California, and their reception by the Natives*, Edward Cavendish Drake, *A new universal collection of authentic...Voyages and Travels*, London, 1768.

Some of the ship's crew ventured ashore on the nineteenth on bark-logs, for the sea ran too high for them to attempt landing with their boats. The good natured natives leaped into the sea to guide the bark-logs in the best manner they were able; and on their making the shore, the Indians led each of the English up the bank, where was an old man, with a deer's-skin spread on the ground, on which they kneeled before the English, who also kneeling, the Indians wiped the water off the faces of the English with their hands.

This ceremony being ended, each of the sailors, supported by two of the savages, was led through a narrow path, of about a quarter of a mile in length, to their huts, where they were welcomed by the music of a very uncouth instrument, being only two jagged sticks, which an Indian rubbed across each other, and accompanied the noise with a voice still more inharmonious than the sound of the instrument. They then all sat down on the ground, and having been regaled with broiled fish, the Indians attended them back in the same manner with their music. (Drake 1768: 104)

The illustration of this strange welcoming manages to encapsulate the different episodes described in the text: there are the Englishmen sailing into shore on bark-logs, aided by the natives, who dive into the water to guide them. On land they are greeted by an old man kneeling on the deer's-skin rug (its antlers still intact!) and led up a path to the musical entertainment that is being performed in their honour. In the top left, they sit on the ground and, if the billowing smoke is anything to go by, eat their 'broiled fish'. The image seems to re-present the words. Positioned facing the text, it could almost be viewed as its mirror image. Almost. For, although this illustration depicts the incidents described in the words, it also reveals its difference as a pictorial genre. It might attempt to imitate the temporality that Lessing associates with poetry, but it does so through its own spatiality: its status as a static and visual work of art. Drawing on the tradition of narrative painting, all the incidents occur in a single frame, the spectators following the 'story' from the bottom right of the image to the top left. Whereas the text moves from one event to the next according to linguistic strategies, the ordering of words, sentences and paragraphs, the illustration achieves this effect in a specifically visual way: its (albeit crude) use of perspective. The viewer is drawn initially to the foreground with its larger figures and the narrative progresses as the people and objects become smaller and fade into the 'distance'. This plate, therefore, is highly paradoxical: it reproduces the text at the same time as it reveals its difference.

It is this focus on difference that sometimes leads to a privileging of the visual. The eighteenth century has been described as ocularcentric, a culture in which sight was regarded as the primary sense (Jay 1993: 83–147). It was through seeing and observing the world that one could come to an understanding of how it functioned. In a period that saw the publication of Isaac Newton's *Opticks* (1704), the circulation of Bishop Berkeley's idea that nothing exists apart from its perception, and Edmund Burke's formulation of the visual categories of the beautiful and sublime, it is no surprise that pictures were often seen as superior to words. In the case of the illustrated travel account, the text could sometimes appear superfluous, books being sold on the strength of their images. When William Hodges turned his attention to India, he commented that his textual observations about the country were subsidiary to the pictures, serving only for amusement and 'to explain to my friends a number of drawings which I had made during my residence in India' (Hodges 1793: iv). The illustrations that accompany Thomas Anburey's *Travels through the Interior Parts of America* are regarded by the author as necessary to a rendering of the scene: 'In order that you may form a just idea of this important place, I have enclosed you a drawing of it', Anburey writes to his anonymous 'friend' (Anburey 1789: 136); 'But that you may more fully comprehend the construction of these unusual fortifications, I have inclosed a drawing' (Anburey 1789: 138).

Anburey's insistence that images are able to show things more 'fully' or 'justly' than words suggests how significant pictures had become in the travel account. Indeed, one might argue that they played a part in constructing tourism as a specifically visual experience, anticipating a culture where, in the words of Susan Sontag, it 'seems positively unnatural to travel for pleasure without taking a camera along' (Sontag 1977: 9). Here are sights to be seen, cultural curiosities to be observed. The diversity of pictures these books contained, along with the common use of large folded plates, made them truly spectacular. Viewers were encouraged to enter this space, repeating, in the very act of spectatorship, the activity of the explorer. But however seductive or appealing the image, this visual engagement was always at odds with itself. The process of viewing was tempered with an awareness of the viewers' own difference from the pictures set before them. This was a voyage, a journey into the unknown, but the distance traversed was ideological as well as geographical.

Self and Other

The gap between viewer and viewed is not simply opened up by the act of reading the eighteenth-century illustrated travel book. It is there in the images themselves and in the physical and analytical distance required to picture a scene. As spectator, the artist is both divorced from the spectacle and an integral part of it, bound up in a complex set of visual power relations.

This was a culture that recognised and exploited the power of the gaze, its ability to control and discipline whatever lay within the field of vision. In 1791 the British philosopher Jeremy Bentham proposed the building of a 'panopticon', a model prison where solitary cells were placed around a central watchtower, which allowed the inmates to be seen at all times, although they could not tell if they were being observed. According to Bentham, it was the very suggestion of surveillance that modified behaviour. And by keeping a watchful eye on the inmates, knowledge could be acquired about them (Bentham 1791).

The panopticon might seem far removed from the illustrations that appeared in contemporary travel books, but these images were also designed to circulate information about the people and places pictured. And with knowledge comes power. It is little wonder that on government-sponsored expeditions artists were commissioned and regulations put in place as to what should be drawn or painted. When the French admiral Jean Francois Galaup (Comte de Lapérouse) set off on his travels he was instructed to direct the draughtsmen to:

> take drawings of all the remarkable land views and situations, portraits of the natives, their peculiar dress, ceremonies, pastimes, edifices, boats, all the land and sea productions of the three kingdoms, if drawings of the objects appear ... likely to facilitate the comprehension of descriptions given by the scientific gentlemen on board. (Galaup 1798: lxxii)

The stern warning to artists to hand over *all* their drawings at the end of the voyage suggests just how important these images were deemed to be.

Hodges also took care to represent the most significant scenes of a voyage. An encounter with the natives was particularly picture-worthy. A family in Dusky Bay, New Zealand, gave the artist the name 'Toe-toe', which Cook took to mean 'marking' or 'painting', when Hodges drew them on his visits (Cook 1777: 75). The published engraving of one of these drawings sets up the hierarchy of visual relations inolved in looking and picturing (see illustration 4.2). Despite the foregrounding of the natives, there is a sense in which it is the invisible spectator who is the focus of this image. All the features of the scene – the family, the curve of the landscape behind them – present themselves to the viewer. The scene is staged for us and we are masters of a vision that seems to show everything. But, of course, this is only ever an illusion. The aspects of the image converge on our sight lines not because we are all-seeing, but because in order to view the engraving we have to occupy a fixed position in front of it, the position once occupied by the artist. The image, then, literally puts the viewers in their place.

Julia Thomas

The Dusky Bay family is positioned in another way, although this is also determined by visual politics. As the spectator is physically distanced from the scene, his or her ideological detachment is intensified. Images like this engrain the imperial attitudes that can also be detected in the texts. There is an attempt both to appropriate the native and to stress difference. The natives here are 'other' and it is the representation itself that constructs this otherness, often by blurring those definitions that are central to European culture. Although the status of the two erect figures seems straightforward, with the patriarch on the right and his wife on the left, the position, and even gender, of the seated figure remains unclear. Is this a male or female? The clothes give nothing away, although the baby being carried suggests that this is a woman. And if she is female, what is her relation to the other figures in the picture? Is she a daughter or another spouse?

It is the ambiguous status of this figure that adds to the otherness of the natives. They are different because they do not fit comfortably into the Western idea of the family and it is the constant allusions to the 'norm' that stress this difference. Not only are the people here labelled a 'family', but the central man and woman are depicted in a classical way with their attitudes and physical appearance pointing to their gender roles: the man is taller than the female, strongly built, and stares admiringly at his partner, while she, with some elegance, drapes herself round a pole and looks shyly at the viewer. The text is similarly concerned to define this family in European terms, although the tentative language ('as we supposed', 'she seemed') betrays uncertainties and anxieties:

> [the family] consisted of the man, his two wives (as we supposed), the young woman before mentioned, a boy about fourteen years old, and three small children, the youngest of which was at the breast. They were all well-looking, except one woman, who had a large wen on her upper lip, which made her look disagreeable; and she seemed, on that account, to be in a great measure neglected by the man. (Cook 1777: 75)

It is this tendency to normalise and homogenise, to make things seem the same the world over, that, paradoxically, emphasises difference. Difference, after all, can only be constituted if it has something to compare itself with, something that it is not. The family in Hodges' picture is defined by its otherness from the European family, even as it is represented in terms of this ideal.

With difference so precariously constructed, the gap between spectator and spectacle, traveller and local inhabitant, also fails to stay in place. In 1794 Dean Mahomet from Patna in Bengal published a two-Volume account of his travels through India while in the service of the East India Company. As an

Indian writing about India for a British audience, Mahomet slips between different subject positions: he is native and foreigner, the observed and the observer; he has a dual identity that implies that difference is not tied to an individual: the 'other' can be within the same. This is similarly suggested in the illustrations that accompanied travel books, where the tourist was on display, as much the object of the viewer's gaze as the local inhabitants. The frontispiece for a collection of voyages published in 1760 (Anon. 1760) shows the 'surprize' of two visitors, who are shown some 'stupendous Ruins' by a local guide, but it is their theatrical gestures and ostentatious garments that makes them the focus of the image. The ruins, however magnificent, have to compete with these travellers for the viewer's attention, and they have little chance of succeeding, for, as their postures and finery indicate, the tourists desperately want to be seen.

It is therefore not simply the natives who signify otherness in eighteenth-century illustrated travel books. The travellers themselves can appear strange and alien. The pictures, as well as showing the unfamiliar, work by defamiliarising what is recognisable. But if difference is not tied to a place or subject, then where is it located? Perhaps it is in the very frock coats and tricorn hats worn by these conspicuous tourists. For one of the primary visual signifiers of otherness in illustrated travel books is clothing itself. Or lack of clothing. The sailors who arrive with Captain Rogers on the coast of California are met by the conventional image of the semi-clad native with feathers in his hair and only a grass skirt to protect his modesty (see illustration 4.3). In fact, it is the costume of these inhabitants that marks the image's deviation from a text that it is otherwise at pains to represent. In the narrative, the natives are described as 'quite naked', with just the women wearing petticoats to the knee (Drake 1768: 104). The picture, however, dresses them up. Only one man is actually nude, although he is shown diving into the sea in a way that keeps his dignity intact.

The garments worn by the inhabitants in this engraving bring to the fore the difference of images from texts: it is acceptable to write that someone is 'naked', but to actually show it is problematic. Pictures have the power to shock the senses. But there is another difference that is at stake here, for, in his grass skirt and feathers, the Californian native wears and perpetuates the myth of primitiveness, savagery, sexual liberty (and immorality) and inferiority. It might also be the case that, dressed in this way, the natives appear more comic. Any threat posed by their muscular bodies is immediately counteracted by their choice of outfits. They are to be laughed at, not feared.

In the illustrations for Drake's voyages, however, the travellers cannot take themselves too seriously either. An account of Admiral Anson's invasion of Paita in Peru, which resulted in his men stealing and dressing up in the clothes of the town's inhabitants, is accompanied by an engraving of

Hogarthian proportions (Drake 1768: 100). As the sailors riotously greet their lieutenant, they are shown in the most astonishing clothes. The leader of the pack sports a woman's dress and a garment is waved in the air as a symbol of their victory. Despite the comedy of this image, it does wear a more sinister aspect. Seizing and dressing up in someone else's clothes is an act of appropriation, a stripping away of the identity of the victim. But what is also suggested here is a carnivalesque overturning of distinctions, the idea that gender, class and racial difference are not fixed but are themselves 'costumes' to be put on or taken off. The sailors who proudly wear the garments of the local inhabitants display both their power over them and the instability of the structures on which this power depends.

Perhaps the illustrations that adorn the pages of eighteenth-century travel books could be seen as similarly contradictory. These images construct difference, setting up distinctions between fact and fiction, word and image, self and other, but they also contain the possibility of subversion, exposing the fact that these categories are far from stable. The task of the angel, who witnesses and interprets the foreign sights pictured in Le Bruyn's frontispiece, might not be so easy after all.

Julia Thomas is Professor of English Literature at Cardiff University, UK, where she specialises in illustration and the relationship between word and image. She has published extensively in these fields, including her latest monograph, *Nineteenth-Century Illustration and the Digital* (2017). Thomas is the Director of *The Illustration Archive*, the world's largest online resource dedicated to book illustrations.

References

Anon. (1760) *A New and Complete Collection of Voyages and Travels ... Illustrated with fifty-two elegant copper plates*, London.
Anburey, T. (1789) *Travels through the Interior Parts of America*, 2 Vols. Vol. 1, London.
Baudrillard, J. (1994) *Simulacra and Simulation*, trans. Sheila Faria Glaser, Michigan: University of Michigan Press.
Benjamin, W. (1968) 'The Work of Art in the Age of Mechanical Reproduction',in Hannah Arendt (ed.), *Illuminations*, trans. Harry Zohn, London: Fontana, pp. 211–244.
Bentham, J. (1791) *Panopticon: or, the Inspection House, containing the idea of a new principle of construction applicable to any sort of establishment, in which persons of any description are kept under inspection*, Dublin: Thomas Byrne.
Le Bruyn, C. (1737) *Travels into Muscovy, Persia and Part of the East-Indies*, 2 Vols. Vol. 1, London.
Le Comte, L.D. (1739) *A Compleat History of the Empire of China*, London.
Cook, J. (1777) *A Voyage towards the South Pole, and round the World ... in the years 1772, 1773, 1774, and 1775*, 2 Vols. Vol. 1, London.

Dalrymple, A. (1773) *A letter from Mr. Dalrymple to Dr. Hawkesworth, occasioned by some groundless and illiberal imputations in his account of the late Voyages to the South*, London.

Drake, E.C. (1768) *A new universal collection of authentic ... Voyages and Travels*, London.

Galaup, J.F., Count de Lapérouse (1798) *The voyage ... round the world in ... 1785, 1786, 1787 and 1788 ... Arranged by M.L.A. Milet Mureau ... Translated from the French*, 2 Vols. Vol. 1. London.

Gilpin, W. (1800) *Observations on the River Wye, And Several Parts of South Wales, &c. Relative Chiefly to Picturesque Beauty: Made in the Year 1770*, London.

Hawkesworth, J. (1773) *An account of the voyages undertaken by the order of His present Majesty for making discoveries in the southern hemisphere*, 3 Vols. Vol. 1, London.

Hodges, W. (1793) *Travels in India, during the years 1780–1783*, London.

Jay, M. (1993) *Downcast Eyes: The Denigration of Vision in Twentieth-Century French Thought*, Berkeley: University of California Press.

Joppien, R. and Smith, B. (1985) *The Art of Captain Cook's Voyages*, Vol.1, New Haven and London: Yale University Press.

Keysler, J.G. (1760) *Travels through Germany ... Illustrated with copper plates, engraved from drawings taken on the spot*, 4 Vols. Vol.1, London.

Lessing, G.E. (1766) *Laocoon; or, on the Limits of Painting and Poetry*, London.

Mahomet, D. (1794) *The Travels of Dean Mahomet* Cork.

Raspe, R.E. (1793) *Gulliver Revived: or, The Vice of Lying Properly Exposed ... by Baron Munchausen ... with Twenty Explanatory Engravings*, London.

Sonnini, C.S. (1799) *Travels in Upper and Lower Egypt*, Vol.1, London.

Sontag, S. (1977) *On Photography*, Harmondsworth: Penguin.

Sparrman, A. (1785) *A Voyage to the Cape of Good Hope, towards the Antarctic Polar Circle, and round the World ... from the year 1772 to 1776*, Vol.1, London.

Stafford, B.M. (1984) *Voyage into Substance: Art, Science, Nature, and the llustrated Travel Account, 1760–1840*, Massachusetts: MIT.

Swift, J. (1727) *Gulliver's Travels* ed. Christopher Fox (1995), Boston and New York: St. Martin's.

Chapter 5

A BEGINNING, TWO ENDS, AND A THICKENED MIDDLE
Journeys in Afghanistan from Byron to Hosseini

Graham Huggan

What kind of travel destination is Afghanistan today? The most recent edition of *Lonely Planet Afghanistan* ("Introducing Afghanistan" 2013) contents itself with stating the obvious: "By any stretch of the imagination, Afghanistan isn't the simplest country to travel in. It's a country recovering from nearly three decades of war, with a host of continuing problems. ... For the visitor, it's a world away from backpacking in Thailand or island-hopping in Greece." Not that the intrepid traveler should be put off, for "with the right preparations, and a constant ear to the ground ... travel in Afghanistan is not only [possible] but also incredibly rewarding." "It's an addictive country to visit, [in which] the history get[s] in your blood and promise[s] to draw you"; meanwhile, "the post-Taliban scene [*sic*] has brought investment to the country for the first time in years."

The manufactured "addictiveness" of Afghanistan is one of the underlying subjects of this article; so too is continuing "investment," as much symbolic as material, in a country that has provided several generations of adventure-seeking travelers with a rich storehouse of heavily mediated images that often bear little resemblance to the realities they experience on the ground. So little resemblance, in fact, that the ground beneath Afghanistan has latterly begun to shift, giving way to the polymorphous perversity by which a proud Central Asian nation, previously assimilated to the (British-Russian) Great Game, now finds itself incorporated all over again, its physical geography paranoically skewed to meet the (American) ideological requirements of a "Greater Middle East." This article raises the following questions: To what extent have travel writing and other related tourist forms been complicit in this shift, the geopolitical coordinates of which closely map onto the "War on Terror"? And to what degree has the "Middle East-ification" of Afghanistan been either challenged or confirmed in both recent and earlier representations of a country which, al-

---

Notes for this chapter begin on page 81.

though it has been envisioned as many things, has rarely been allowed to be itself? Finally, the article explores the triangulated relationship between travel, empire, and colonial modernity in three loosely connected quest narratives set in and around Afghanistan, only the first of which conforms to most conventional understandings of travel writing, and which range from the mid twentieth century (Robert Byron's *The Road to Oxiana*) to the early twenty-first (Khaled Hosseini's *The Kite Runner*, Michael Winterbottom's *In This World*).

Most contemporary representations of Afghanistan tend to alternate between negative visions of the country as a failed state, a land in ruins and, above all, a theater of war in which lives on both sides (but especially the Euro-American side) are being tragically lost. The military figures certainly make for grim reading. The number of American fatalities, over a decade into Operation Enduring Freedom, is now well over the two thousand mark. The UK has also lost more than four hundred soldiers in what an increasing number of commentators are seeing as an "un-winnable" battle against the politically ousted but still resilient Taliban, whose ragtag insurgents consist mainly of Afghans, although Arab and Uzbek fighters are also said to be involved.[1] Needless to say, the figures on the Afghan side, both military and civilian, are significantly higher than this, and—as in the Iraq conflict—the true number of casualties will probably never be known. As Derek Gregory asserts, some of the killing has been—to put it mildly—indiscriminate, while substantial fatalities have been sustained as an indirect consequence of intervention, including those thousands who have died "through the secondary effects of targeting civilian infrastructure" or when "relief columns from international aid agencies were halted or delayed" (2004: 70).

Conflicts of this kind, devastating as they are, are hardly new to a country that has undergone a long and painful history of domestic strife and foreign incursion (Gregory 2004: 30). It was Lord Curzon who once famously described Afghanistan, not as conjuring up a sense of moribund romance or mystical remoteness but as contributing one of the crucial pieces to a chessboard on which a game for the domination of the world was being played out (quoted in Gregory 2004: 31). Given this context, it is understandable that a powerful aura of political brinkmanship still hovers over travel writing to the region. Travelogues about Afghanistan over the last couple of centuries have tended to present the country in terms of a particular set of physical and psychological challenges to the traveler. However, this has not necessarily diminished the repetitive and indeed predictable quality of the writing, which though perhaps less prone to cliché than some of its nineteenth- and twentieth-century Orientalist counterparts, still circles persistently around a series of readily identifiable images and tropes (Fowler 2007: 3, 9–10; see also Featherstone 2005). Corinne Fowler, in her excellent overview of nineteenth- and twentieth-century verbal and visual representations of Afghanistan produced by British travel writers and, more recently, newspaper and television journalists, points out two prevalent sets of image-making processes. These include those processes of *medievalization* by which Afghanistan is presented as being at a "pre-civilised stage of social and cultural development"; and those processes of *Wild Westification* by which Afghanistan, an apparently lawless place whose inhabitants seem almost pathologically predisposed to violence, becomes a latter-day version of the Badlands, a "Wild West in the East" (Fowler 2007: 61).

This last epithet raises a question that remains centrally important to mediated understandings of Afghanistan and its people in the Western cultural, political, and economic contexts. *Where* is Afghanistan? This question, as one might suspect, turns out to be less geographical than ideological. Or, perhaps better, it turns out to be about what Gregory calls —following Said— "imaginative geographies" of Afghanistan as a recurring site of "uncivil war and transnational terrorism," in which the distinctive topographical and tropological markings of a colonial present map their coordinates onto the cultural-political atlas of the colonial past (Gregory 2004: 71; see also Said 1978). Certainly, Gregory's eponymous book, *The Colonial Present* (2004), makes little attempt to justify the huge geographical distances that separate his three primary case studies—Afghanistan, Iraq, and Palestine—leaving it to his readers to infer that these three modern theatres of conflict are comparable, explicitly or implicitly, within the context of an ideologically polarized if increasingly deterritorialized and culturally/economically globalized world.

This world, Gregory insists, needs new forms of cognitive mapping that complicate Jameson's heady notion that new, cybernetic forms of information transfer can be used to "suppress the space that held the colony apart from the metropolis in the modern period" (quoted in Gregory 2004: 11–12). On the contrary, Gregory says, cognitive mappings of this kind impose their own unequal and uneven geographies; and, similarly, celebratory visions of a "shrinking world" or the "global village" have obscured the ways in which distances between countries, cultures, and peoples can significantly expand (252). Indeed, one of the features of the modern globalized world is what Gregory calls "the proliferating partitions of colonial modernity" (252)—the means, both material and symbolic, by which those ideologically differentiated spaces and places of colonial conflict that operate under the code words "Afghanistan," "Iraq," and "Palestine" have been strategically refashioned by contemporary imaginative geographies performed, in particular, in the aftermath of the 2001 terrorist attacks in Washington, DC and New York (19).

A related version of the same thesis is provided by another North American-based cultural geographer, Dona Stewart, who—in critically analyzing the rationale behind the Bush administration's Middle East policy—uncovers a "very specific geographical conceptualisation of the region, formed in the aftermath of the 9/11 attacks" (2005: 400). Prior to the attacks, says Stewart, US Middle East policy had placed emphasis on the stability, both political and economic, of countries in the region whatever their level of democratization or civil society participation; however, 9/11 changed this. The post–9/11 approach has generally been (1) to pursue terrorists and the regimes that support them, and (2) to facilitate the democratic transformation of governments in the region—a process understood "common-sensically" to require the removal of the Taliban from Afghanistan and Saddam Hussein from Iraq. As Stewart suggests, the George W. Bush administration (and, although her article predates this, the Obama administration that has since followed it) have struggled to define the geographical parameters of the Middle Eastern region, and a series of policy initiatives, increasingly clustered around the "War on Terror," have emerged as vehicles for producing political reform in a region variously described as the "Greater" or "Broader" Middle East. The current working definition of the "Greater Middle East," for example, contains no fewer than twenty-six nations: the twenty-two nations of the Arab world, plus Turkey, Israel, Pakistan—and Afghanistan (2005: 401).[2] (That Afghanistan is not geographically in the Middle East at all will need no further reminder in this article, though its imaginative and ideological coordinates are "Middle Eastified" in at least some of the examples I present below.)

Stewart is not as open as Gregory in attributing this "thickening" of the Middle East to US imperial ambition. However, she is critical of the turn to policy initiatives which, in "attempting to generalize a [cultural] and geographical landscape full of contradictions ... have minimized a nuanced understanding of [cultural, economic, and political] diversity within the region," at worst falling into the kinds of reductionism that justify a one-size-fits-all approach (2005: 402). She is similarly critical of US and US-backed attempts to strong-arm the "progress to democracy" in a region identified as suffering from collective "democracy deficit." (The George W. Bush regime in particular repeatedly cited this as the principal explanatory factor in the rise of terrorist groups like al-Qaeda. The Bush administration argued that the perceived lack of civil society and political freedom in the region had not only threatened peace, prosperity, and modernity, but had also shaped an authoritarian environment in which al-Qaeda and other rogue insurgencies could create and consolidate support [Stewart 2005: 405].) Recent US Middle East policy, Stewart concludes, has fomented the very unrest it was originally intended to neutralize; moreover, it is open to charges of hypocrisy insofar as it continues to prop up regimes, such as those in Saudi Arabia and Pakistan, which arguably have little interest in pushing for democratic reform (2005: 415; see also Stewart 2012).

Graham Huggan

Although I am in broad agreement with Gregory and Stewart, it is not my intention here to debate the finer points of US and/or Coalition administrative and military policy. Rather, I want to use the idea of a "thickened" Middle East that both of these scholars either explicitly (Stewart) or implicitly (Gregory) engage with as an informing backdrop to my commentary on twentieth- and early twenty-first century travel writing's contributory role. To describe travel writing as "contributing" to the current situation in Afghanistan is, of course, to run the risk of being seen as either irritatingly moralistic or hilariously ingenuous. But as Gregory—who has himself written illuminatingly on travel narrative—points out toward the end of *The Colonial Present*, travel writing is part of those modern-day cultures of travel and the tourist industries that support them, in which the privileges of mobility are at least potentially complicit in the social and economic inequalities they suppress (2004: 256–257). Travel writing, after all, even as it morphs into the new conditions of global travel, provides a reminder—as Homi Bhabha puts it—that "the globe shrinks for those who own it, [but not] for the displaced or the dispossessed" (quoted in Urry 2002a: 62).[3]

In what follows, I look at one "beginning" and two "ends," between which Afghanistan's geographical misplacement in the Middle East serves as a strategically thickened "middle," and through which I aim to track British and American ideological investments in Afghanistan as disaster zone, "adventurous location" (Fowler 2007: 9), and over-determined political battleground for conflicting versions and understandings of both the West (America, Europe) and the East (Russia, Asia). I make no apologies for working here—in the last two cases at least—with examples that would not satisfy most conventional definitions of travel writing. For although my first example is Robert Byron's 1937 classic *The Road to Oxiana*, long since adopted into the canon of modern British travel literature, my second and third examples are Khaled Hosseini's 2003 novel, *The Kite Runner*, and—stretching things still further—*In This World*, Michael Winterbottom's 2002 faux-documentary film. Without dwelling on it, my rationale for opening things up this way is essentially the same as in my latest book on travel writing. The book argues for a radically expanded view of the "travel text" that imbricates travel writing with other forms of travel practice, showing in the process that modern travelogues, "as cultural documents of the present, are bound up in all kinds of movements that narrower, if still expansive, definitions of travel writing as 'factual fictions' (Holland and Huggan) or 'sub-species of memoir' (Fussell) have been unable to embrace" (Huggan 2009: 5).

After this lengthy preamble, let me begin at the "beginning" with Robert Byron's *The Road to Oxiana* ([1937] 2010). Byron's work is generally and justifiably acknowledged to be a foundational text of modern British travel writing, though not all readers might go as far as Paul Fussell, who elevates it to the heights of *Ulysses* and *The Waste Land*, crowning the hagiography by stating that "if the souls of the dead could come and assist us, [Byron] would be venerated as the saint of all whose imaginations come alight at the thought of travel in the now obsolete sense" (1980: 79). Fussell duly assimilates Byron to his unashamedly nostalgic view of interwar British travel writing, which he sees as being motivated by an idiosyncratic mixture of the stylistically modernist and the temperamentally anti-modern—qualities apparently embodied in the anachronistic figure of Byron himself.

This last view of Byron, and of Byron's writing, is only accurate up to a point. As Howard Booth remarks in one of relatively few critical essays on Byron, *The Road to Oxiana* is very much of its time even though its author prefers to pontificate on earlier historical periods. Leavened by the kind of pseudo-aristocratic high jinks for which British travel writing would later become known, the text is deadly serious about the understated elegance of Afghan, Persian, and Turkmen architecture. It is just as serious in its visible apprehension about the shadows being cast by nationalism, xenophobia, and intolerance across interwar Europe, at least some of which Byron correctly attributes to Europe's weakening position in a late colonial world (Booth 2002: 101). As Booth persuasively suggests, Byron is much more than Fussell's sycophantic vision of an overconfident gentleman-traveler in a three-piece suit, expatiating on the wonders of non-Western architecture; but much less in view of the considerable anxieties and intermittent awkward silences that punctuate *The Road to Oxiana* as a late colonial text. Booth correspondingly sees evidence in the text for the consolidation as well as the condemnation of colonialism at a time when being British in the Middle East—the last frontier of British colonial expansion—was far from an easy nationalist enterprise; and when there was increased questioning, both in the metropolis and the colonies, as to whether the British Empire could be sustained.

Byron's irascibility, much in evidence throughout the text, can be partly attributed to this lurking recognition of unease, which he sometimes allays by deflecting onto other travelers—notably tourists, whom he disdains, and adventurer-explorers, whom he sees as little more than "overgrown prefects and pseudo-scientific bores despatched by congregations of extinguished officials to see whether sand-dunes sing or snow is cold" ([1937] 2010: 317). Over and against these cosseted buffoons, Byron posits the ideal of the physically hardened but culturally sensitive "literary traveller," who takes genuine enjoyment in an appreciation of other cultures; and who, "while he may or not be naturally observant," at least refuses to "dress up the result in thrills that never happened or science no deeper than its own jargon," trusting instead the evidence of his own senses and using up "what eyes he [has]"

([1937] 2010: 318). As Booth suggests, tirades like these belong to a tradition of self-consciously "literary" travel writing that celebrates the "possibilities for cultural renaissance and personal growth that ... result from cross-cultural encounters"—a liberal-humanist tradition that confronts its own limits in the privileges accorded to the enlightened British traveler, who is less than keen to see that his mobility around the world has come at the expense of others, as well as in the residual colonial habits of mind that lurk behind, say, Byron's violently anti-Arab sentiments, which allow him to support the idea of a reformed Empire even as he flays the arrogance and prejudice of the British ruling class (Booth 2002: 110).

As always with foundational texts, it is possible to challenge the "beginnings" that works like *The Road to Oxiana* are said to initiate.[4] Certainly Byron's work is indebted to its own precursors, not least those eighteenth- and nineteenth-century scholars of the East who, sometimes preferring the pleasures of connoisseurship to more rigorous scholastic activities, were instrumental in performing what Said (1978) calls the "curatorial" mode of Orientalism that remains influential to the present day.[5] A rather different mode is that performed by the contemporary "Oriental" *tourist novel*, not discussed in Fowler's otherwise comprehensive 2007 survey of images of Afghanistan in nineteenth- and twentieth-century writing, which freely admits that even the most factual of travel narratives can be based on fictional premises, and which acknowledges the artificiality of separating travel writing from other forms of popular knowledge production about Afghanistan in both nineteenth-century cultural imaginaries of "Asia" and their twentieth- and early twenty-first-century counterparts, including post–9/11 imaginative geographies of the "Middle East."

Afghan American writer Khaled Hosseini's hugely popular novel *The Kite Runner* (2003), which held a spot on the *New York Times* bestseller list for more than five years, can be used in this last sense to mark the uneasy transition from travelogues like *The Road to Oxiana*, which function at least in part as "factual fictions" (Holland and Huggan 1998), to tourist novels which, however obvious their fictional status, seem always to run the risk of being taken for fact. The genre of the "tourist novel," despite its widespread popular influence, remains relatively untheorized. Sarah Brouillette has associated tourist novels with "postcolonial guilt," the process by which local cultural material produced for mass consumption in the global literary marketplace generates forms of "anxious responsibility" on the part of both writers and readers, who share the fear—though perhaps for different reasons—that global currencies of suffering may turn autoethnographic fiction into "compassionate fodder for the travel book" (Walcott, quoted in Brouillette 2007: 6). True to form, *The Kite Runner* has generated self-conscious criticism of its own strategically exoticizing tendencies, for example via Hosseini's apparent willingness to be seen as a "representative" Afghan author, or via the novel's only partly self-ironizing capacity as a literary tour guide, autoethnographic in inspiration, to the country it fictionally presents (Jeferess 2009; White 2009).

Perhaps the most interesting reading of *The Kite Runner* to date is by Mandala White. White sees the novel as a market-conscious cross between tourist novel, exotic romance, and Western-oriented liberal-humanist fable, and uses it to open up a debate around the figure of the sympathetic reader, who guiltily consumes the "dark touristic" atrocities—stonings, wife beatings, starvation, orphan paedophilia—of which the novel repeatedly relates (White 2009: 17; see also Lennon and Foley 2000).[6] The novel's journey to redemption, White suggests, is similarly projected onto the figure of its implied (non-Afghan) reader, who consents willingly to be guided around war-torn Afghanistan, is duly horrified by what she witnesses, and is rhetorically though not necessarily ideologically convinced by Hosseini's melodramatic representation of Afghans as "victims of war [with] America as their saviour" (White 2009: 27). For example, protagonist Amir, after having redeemed his earlier misdemeanors by returning riskily to a Taliban-tormented Afghanistan, is duly rewarded by being granted his new life as a "lucky migrant" to the United States.

Key to this reading is Amir's status as a "semi-insider," Rob Nixon's useful term for privileged postcolonial writers who have an obvious experiential connection to the places they delineate, but whose relationship to these places is more like that of a "tourist" than that of a "native" or, perhaps better, is a hybrid combination of these always-already entangled positionalities at once (Nixon 1992; see also Bruner 2005; White 2009). Seen this way, Amir serves in the novel as a double stand-in: for the *writer* (Hosseini), who has lived most of his life in a country other than that for which he still seems willing to be accepted as a representative; and for the *reader*, who is encouraged to identify sympathetically with the journey—part estranging, part familiarizing—that allows Amir to rediscover the authentic "Afghan-ness" he already carries in his blood (White 2009: 45–46). These substitutions are replicated at several different levels, so that the novel comes to play between virtually interchangeable positions—now "tourist," now "tour guide"; now "participant observer," now "native informant"; now "marginalized local," now "privileged foreigner"—depending on the perspective from which it is read. The end effect, White convincingly suggests, is to create a pseudo-ethnographic account of Afghanistan that both plays to postcolonial guilt and works toward its absolution through the figure of the sympathetic reader. Thus, when "Amir confronts his privilege through his interaction with 'exotic' Afghanistan," and the reader "relativises her life through witnessing the suffering of Hosseini's characters at the hands of the Taliban," we are given to understand that Amir's guilt, which is eventually absolved, is potentially our own (2009: 28).

There are numerous problems, of course, with this liberal-humanist self-conception of the novel. First, as White suggests, it offers sympathy as a way out for readers whose emotional engagement simultaneously supposes a safe degree of distance from the country. Second, its representations of both the Afghan present and the Afghan past are largely based on marketable, paradoxically reassuring First World stereotypes—the victimized ethnic-minority boy (Hassan), the put-upon Third World woman (Soraya)—which circulate, in appropriately spectacular fashion, as commodities within a symbolic economy of violence, suffering, and loss. And third, it presents

> an exoticised image of Afghanistan that sits easily alongside the official rhetoric and military response to the Middle East in post 9/11 America. Amir emerges from his 'dark tourist' experience with a melodramatic understanding of the country he has just encountered: Afghanistan for him is a land of Good Muslims and Bad Muslims, and the latter need outside forces to save them from the former. (White 2009: 27)

Two brief comments can be added to this. The first comes via Mahmood Mamdani whose "Good Muslim/Bad Muslim" dichotomy—also invoked by White—mobilizes another powerful critique of *The Kite Runner*. For Mamdani, fundamentalist Islam emerges at the onset of modernity: symbiotically entwined with what it opposes, it is as modern as modernity itself (Mamdani 2004: 45; see also White 2009: 51). By consigning the Taliban to a stereotypically medievalized past (which, as we have seen, is fully in keeping with a well-established pattern in literary representations of the country), Hosseini lends involuntary support to the "civilizationist" rhetoric that positions a superseded Afghan culture against its American liberal-progressive counterpart (White 2009: 51). This rhetoric, in driving a wedge between religious and political Islam, notably fails to recognize that radical Islam is a *political* response to US-style "turbo-capitalist" imperialism, which—as David Harvey (2003) has amply demonstrated—feeds voraciously off the economic and political divisions it continually recreates (White 2009: 51–54).

Implicit here is that countries like Afghanistan need to modernize, but in American terms—which brings me to my second comment. It is unclear whether White's implied positioning of Afghanistan in the Middle East is a category mistake, but even so it is certainly instructive. *The Kite Runner*, in its defense, is not so crude as to absorb Afghanistan into the "Muslim world" (e.g., via popularized conceptions of a "Greater Middle East") or to homogenize Afghan culture. To some extent at least, the different representational levels at which the novel operates militate against, or at least complicate the ready acceptance of, stereotypical Orientalist images and tropes. However, nowhere does the novel suggest that Afghanistan's plight is connected to the global crisis of colonial modernity, while the novel's touristic discourse, translating across cultures, is deliberately designed to create a distancing effect. Hosseini's Afghans are thus "humanized" *at the same time as* they are culturally and politically differentiated, and his Afghanistan ideologically "Middle East-ified" so as to create the false dichotomies—between West and rest, self and other, past and present—that its colonial modernity, fully coeval with America's, contests.

Although this might seem an appropriate place to end, I want to supply—for reasons that will soon become apparent—a second ending, and will therefore close with a few brief comments on Michael Winterbottom's award-winning docudrama, *In This World* (2002). These comments can be gathered here in three overlapping categories. *Diegetically*, the film follows the fortunes of two Afghan refugees as they embark on a dangerous overland trip from their makeshift camp in northwest Pakistan. The trip, arranged by disreputable people smugglers (cf. Karim in *The Kite Runner*), takes its increasingly desperate convoy across a series of bleak post–9/11 securitized landscapes—the checkpoint-dotted steppes and deserts of Baluchistan and Iran, the closely patrolled mountain range that separates Iran from Turkey—during which time they risk their lives several times in pursuit of their goal, a new life in London. However, of the two, only one, teenage Jamal, makes it—and even then, only temporarily—while the other, his older nephew Enayat, is no longer "in this world." *Generically*, the film is probably best attached to the tradition of observational cinema, a documentary film style that alternates between factual reportage and explicit fictionalization, and which, while it usually focuses on details of lived personal experience, shifts constantly between "private" and "public" worlds (Taylor 1998). *Aesthetically*, it is characterized by a high degree of stylistic hybridity, switching quickly between the conventions of the road movie, (faux-) documentary, and ethnographic film. Code-switching creates a dialectical interplay between familiarity and unfamiliarity so that, at times, we are reassured that the refugee protagonists are just like us—their world is also our world—while, at others, we are shown worlds of experience that are entirely and shockingly unlike our own.

The film raises interesting questions about the relationship between refugees and the moving image (see Wright 2002, especially his discussion of the connections between refugee quest narrative and the road movie). My emphasis here, though, is on enforced migration—the flipside of celebratory travel—which some see as having replaced the outdated idea of travel as freedom with the altogether harsher realities of "indentured trade projects, racialized exclusion, and orientalist cultural work" (Kaur and Hutnyk 1999: 4; see also Huggan 2009). Refugee journeys, as the film makes clear, debunk the myth of a borderless world while simultaneously demonstrating the global reach of a "multi-partitioned" colonial modernity (Gregory 2004). The refugee acts here as an "emblematic figure for globalized modernity's violent displacements and disruptions" (Huggan 2009: 179); and as a sobering reminder not just that the freedom allowed to some is emphatically forbidden to others, but also that freedom for these others may turn out to be an illusory goal. The "end" of Jamal's journey, in this respect, turns out to be just another false beginning: his asylum request in the UK is rejected and though he is granted special leave to enter the country, he is certain to be deported within the next couple of years.

Films like *In This World* can be used to illustrate both the expanding horizons and reinforced limits of the colonial present, caught as it is between globalization's aggressively *deterritoralizing* impulses and colonial modernity's *reterritorializing* imperatives, which "forever install partitions between 'us' and 'them'" (Gregory 2004: 253). This double structure also cuts across the history of modern travel writing which, like the forms of disavowed tourism it instantiates, can be seen as an admittedly unreliable index of modernity itself (MacCannell 1976). Modernity is partly defined by those it excludes—a truth painfully reaffirmed in the current global conjuncture. But in another sense these are *false* exclusions, because—and this is surely one of the several unlearned lessons of colonialism—modernity is shared by those very people whose entirely legitimate claims to be modern are withheld.

At their best, modern cultures of travel—to which travel writing belongs—attest to the desire to open up the world to engagement between different ideas, cultures, and peoples; to what the Indian historian Dipesh Chakrabarty (2002) felicitously calls different "habitations of modernity." Conversely, at their worst they have contributed—and still contribute—to a myopic view that distinguishes, with a nostalgia as damaging as the condescension it counters, between "modern" and "non-modern" or "pre-modern" worlds. This is not to collapse history's different beginnings and ends, but rather to suggest that considerable imaginative as well as material resources will continue to be needed in order to negotiate its own inevitably crowded middle grounds. This requires in turn a capacity—to cite Gregory one last time—to turn "[imperial] imaginative geographies into geographical imaginations that can enlarge and enhance our sense of the world and enable us to situate ourselves within it with care, concern, and humility" (2004: 262). Certainly, as the fatality count rises inexorably in Afghanistan, capacities are required that will have to be far greater than those that have recently been evidenced in George W. Bush's geographically expanded but imaginatively impoverished "Middle East."

A Beginning, Two Ends, and a Thickened Middle

Graham Huggan teaches in the School of English at the University of Leeds, UK. His research spans three fields: postcolonial studies, tourism studies, and environmental humanities. His latest book is *Colonialism, Culture, Whales* (Bloomsbury, 2018), and published work on travel writing includes the 2009 monograph *Extreme Pursuits: Travel/Writing in an Age of Globalization* and, co-authored with Patrick Holland, *Tourists with Typewriters: Critical Reflections on Contemporary Travel Writing* (1998), both with University of Michigan Press.

Notes

1. At the time of this article's writing (late 2013), Operation Enduring Freedom persists, although its official function has recently shifted from that of a "combat" to that of a "support" mission. The distinction is not necessarily as hard and fast as it appears; similarly, the current phased withdrawal of troops from the United States, the UK, and elsewhere is not a sign that the operation will shortly be ending, nor does it imply that the completion of the withdrawal phase (scheduled for 2014) will leave no Coalition troops in Afghanistan. For updated figures on US and other casualties, see the iCasualties website; for other recent information, see the online articles listed in the bibliography.
2. See Stewart (2005); for a more fully updated version of the argument, see Stewart (2012). For contending definitions of the "Middle East"—sometimes conflated with or replaced by the equally problematic terms "Islamic," "Muslim," or "Arab world"—see, for example, Hanafi (2000), Hourani (2005), Lewis (1994, 2001), Mansfield (2013), and Said (1997). A useful recent collection of critical viewpoints is that of Alsultany and Shohat (2013).
3. There is a growing body of secondary literature on travel writing and globalization: two recent book-length examples are Lisle (2006) and Huggan (2009).
4. Earlier contenders might include Mountstuart Elphinstone's *Account of the Kingdom of Caubul* ([1815] 1972) and Lady Florentia Sale's *Journal of the First Afghan War* ([1843] 2003), although both might afoul of modern (twentieth-century) definitions of travel writing. Sale's narrative, whether it is travel writing or not, is certainly one of the sources behind Kipling's *Kim* (1901), which might itself be seen as a foundational text—one whose fictional status separates it from most conventional histories of travel writing, but whose considerable shadow lurks behind many, possibly even most, of the Afghanistan-based British travel narratives that have been written since, from the early twentieth century up to the present day (Fowler 2007).

    One manufactured tradition of British travel writing about Afghanistan might trace a line from Kipling to Byron to Newby ([1958] 1974) to Chatwin (1998) (with honorable mentions to Elliot and Stewart); while another, female counter-tradition might begin with Sale then jump to Stark ([1970] 2010) and—Irishness notwithstanding—Murphy ([1965] 1995) before ending with, say, Lamb (2002). The ambivalence that Booth finds in Byron's work is arguably integral to both traditions, which Featherstone (2005) among others sees as matching a tendency to cultural cliché with an equally predictable aptitude for seeking the moral high ground (see also Fowler 2007). As Featherstone (2005) suggests, both British and American travel writing about Afghanistan has historically been trapped in forms of moral relativism that are characteristic of the genre as a whole—hence the desire to castigate the moral repugnance of, say, the Taliban versus the agonized awareness that such moral grandstanding is inevitably compromised by the traveler's practical need to escape—both literally and metaphorically—to "higher ground."

Ultimately, however—and as Fowler (2007) seems to concede—the search for a British (or any other national) tradition of travel writing about Afghanistan is likely to prove as elusive as the attempt to separate out "non-fictional" travel narratives from other kinds of travel texts. The inclusion of visual material complicates matters further still. For example, the recent history of US involvement in Afghanistan has produced a spate of films, mostly of dubious value, ranging from high-profile liberal drama (*Lions for Lambs* [2007] and, loose geography permitting, *Zero Dark Thirty* [2012]) to investigative documentary (*Restrepo* [2010]), to ill-advised comedy and horror (*Rock the Kasbah* [forthcoming 2014], *The Objective* [2008]). To travel writing purists (a suspiciously oxymoronic term) most of these films would by definition be excluded from "literary" approaches to travel narrative, yet they share at least some of the ideological proclivities to be found in earlier "non-fictional" travelogues as well as contemporaneous tourist novels, both of which vehicles tend to articulate a particular, often voyeuristic, kind of "tourist gaze" (Urry 2002b).

5. The literature of curatorial Orientalism is vast; this is not the place to rehearse it. For ease of reference, Timothy Mitchell's "Orientalism and the Exhibitionary Order" (originally published under a different title in 1989) still remains one of the classic overview essays. For a more recent overview, see Teo (2013).

6. "Dark tourism," a term whose coinage is generally attributed to the British social scientists John Lennon and Malcolm Foley, refers to the network of social processes by which disaster zones of both past and present are turned into tourist attractions and "late-modern pilgrimage sites" (Lennon and Foley 2000: 3). For Lennon and Foley, dark tourism points to "a fundamental shift in the way in the way death, disaster and atrocity are being handled by those who offer associated tourism products" (2000: 3). This shift can be attributed to the changing circumstances of a late-capitalist world in which death and disaster are routinely turned into commodities "for consumption in a global communications market" (Lennon and Foley 2000: 5; see also Huggan 2009: 101). Fiction participates in this conversion process: for further commentary on the function of *The Kite Runner* as an exemplary "dark tourist novel," see White (2009). For updated research on "dark tourism," sometimes called "thanatourism," see Friedrich and Johnston (2013) and Sharpley and Stone (2009).

References

"Afghanistan: Key Facts and Figures." 2010. BBC, 5 July. http://news.bbc.co.uk/1/hi/uk/8143196.stm (accessed 10 December 2010).

Alsultany, Evelyn, and Ella Shohat, eds. 2013. *Between the Middle East and the Americas: The Cultural Politics of Diaspora*. Ann Arbor: University of Michigan Press.

Bigelow, Kathryn, dir. 2012. *Zero Dark Thirty*. Columbia Pictures.

Booth, Howard J. 2002. "Making the Case for Cross-Cultural Exchange: Robert Byron's *The Road to Oxiana*." Pp. 99–111 in *Cultural Encounters: European Travel Writing in the 1930s*, ed. Charles Burdett and Derek Duncan. New York: Berghahn Books.

"British Troops to Be Pulled Out of Afghanistan." 2013. *Mail Online*, 8 December. http://www.dailymail.co.uk/news/article-124153/British-troops-pulled-Afghanistan.html (accessed 8 December 2013).

Brouillette, Sarah. 2007. *Postcolonial Writers in the Global Literary Marketplace*. Houndmills: Palgrave Macmillan.

Bruner, Edward. 2005. *Culture on Tour: Ethnographies of Travel*. Chicago: University of Chicago Press.

Byron, Robert [1937] 2010. *The Road to Oxiana*. London: Vintage Books.

Chakrabarty, Dipesh. 2002. *Habitations of Modernity: Essays in the Wake of Subaltern Studies*. Chicago: University of Chicago Press.

Chatwin, Bruce. 1998. *What Am I Doing Here.* London: Vintage Books.
"Countdown to Drawdown: 10 Facts about US Withdrawal from Afghanistan." 2013. http://countdowntodrawdown.org/facts.php (accessed 30 July 2013).
Elliot, Jason. 2001. *An Unexpected Light: Travels in Afghanistan.* New York: Picador USA.
Elphinstone, Mountstuart. [1815] 1972. *An Account of the Kingdom of Caubul.* Karachi: Oxford University Press.
Featherstone, Kerry. 2005. "A Problematic Subject: Afghanistan in Two Contemporary Travel Narratives." http://erea.revues.org/609 (accessed 3 December 2013).
Fowler, Corinne. 2007. *Chasing Tales: Travel Writing, Journalism and the History of British Ideas about Afghanistan.* Amsterdam: Rodopi.
Friedrich, Mona, and Tony Johnston. 2013. "Beauty versus Tragedy: Thanatourism and the Memorialisation of the 1994 Rwandan Genocide." *Journal of Tourism and Cultural Change* 11, no. 4: 302–320.
Fussell, Paul. 1980. *Abroad: British Literary Traveling between the Wars.* New York: Oxford University Press.
Gregory, Derek. 2004. *The Colonial Present: Afghanistan, Palestine, Iraq.* Oxford: Blackwell.
Hanafi, Hassan. 2000. "The Middle East, in Whose World?" Pp. 1–9 in *The Middle East in a Globalized World*, ed. B.O. Utvik and K.S. Vikør. Bergen: Nordic Society for Middle East Studies.
Harvey, David. 2003. *The New Imperialism.* New York: Oxford University Press.
Hetherington, Tim, and Sebastian Junger, dirs. 2010. *Restrepo.* Outpost Films.
Holland, Patrick, and Graham Huggan. 1998. *Tourists with Typewriters: Critical Reflections on Contemporary Travel Writing.* Ann Arbor: University of Michigan Press.
Hosseini, Khaled. 2003. *The Kite Runner.* London: Bloomsbury.
Hourani, Albert. 2005. *A History of the Arab Peoples.* London: Faber and Faber.
Huggan, Graham. 2009. *Extreme Pursuits: Travel/Writing in an Age of Globalization.* Ann Arbor: University of Michigan Press.
"iCasualties/Operation Enduring Freedom/Afghanistan." 2013. http://iCasualties/org/oef/ (accessed 26 July 2013).
"Introducing Afghanistan." 2013. http://www.lonelyplanet.com/afghanistan (accessed 1 December 2013).
Jeferess, David. 2009. "To Be Good (Again): *The Kite Runner* as Allegory of Global Ethics." *Journal of Postcolonial Writing* 45, no. 4: 389–400.
Kaur, Raminder, and John Hutnyk, eds. 1999. *Travel Worlds: Journeys in Contemporary Cultural Politics.* London: Zed Books.
Kipling, Rudyard. 1901. *Kim.* London: Macmillan.
Lamb, Christina. 2002. *The Sewing Circles of Herat: My Afghan Years.* London: HarperCollins.
Lennon, John, and Malcolm Foley. 2000. *Dark Tourism: The Attraction of Death and Disaster.* London: Continuum.
Levinson, Barry, dir. 2014 (forthcoming). *Rock the Kasbah.* QED Films.
Lewis, Bernard. 1994. *The Shaping of the Modern Middle East.* New York: Oxford University Press.
Lewis, Bernard. 2001. *The Middle East: 2000 Years of History from the Rise of Christianity to the Present Day.* Beverly Hills, CA: Phoenix.
Lisle, Debbie. 2006. *The Global Politics of Contemporary Travel Writing.* Cambridge: Cambridge University Press.
MacCannell, Dean. 1976. *The Tourist: A New Theory of the Leisure Class.* New York: Schocken.
Mamdani, Mahmood. 2004. *Good Muslim, Bad Muslim: America, the Cold War, and the Roots of Terror.* Johannesburg: Jacana.
Mansfield, Peter (revised and updated by Nicolas Pelham). 2013. *A History of the Middle East.* London: Penguin.
Mitchell, Timothy. 1989. "Orientalism and the Exhibitionary Order." Pp. 289–317 in *Colonialism and Culture*, ed. Nicholas Dirks. Ann Arbor: University of Michigan Press.

Murphy, Dervla. [1965] 1995. *Full Tilt: Dunkirk to Delhi by Bicycle*. London: Flamingo.
Myrick, Daniel, dir. 2008. *The Objective*. Jaz Films.
Newby, Eric. [1958] 1974. *A Short Walk in the Hindu Kush*. London: Picador.
Nixon, Rob. 1992. *London Calling: V.S. Naipaul, Postcolonial Mandarin*. Oxford: Oxford University Press.
Redford, Robert, dir. 2007. *Lions for Lambs*. 20th Century Fox.
Said, Edward W. 1978. *Orientalism*. New York: Vintage Books.
Said, Edward W. 1997. *Covering Islam: How the Media and the Experts Determine How We See the Rest of the World*. New York: Vintage Books.
Sale, Lady Florentia. [1843] 2003. *A Journal of the First Afghan War*. Oxford: Oxford University Press.
Sharpley, R., and P.R. Stone, eds. 2009. *The Darker Side of Travel: The Theory and Practice of Dark Tourism*. Bristol: Channel View.
Stark, Freya. [1970] 2010. *The Minaret of Djam: An Excursion in Afghanistan*. London: I.B. Tauris.
Stewart, Dona. 2005. "The Greater Middle East and Reform in the Bush Administration's Ideological Imagination." *Geographical Review* 95, no. 3: 400–424.
Stewart, Dona. 2012. *The Middle East Today: Political, Geographical, and Cultural Perspectives*. New York: Routledge.
Stewart, Rory. 2004. *The Places In Between*. London: Picador.
Taylor, Lucien. 1998. "Introduction." Pp. 3–21 in David McDougall, *Transcultural Cinema*. Princeton: Princeton University Press.
Teo, Hsu-Ming. 2013. "Orientalism: An Overview." *Australian Humanities Review* 54. http://www.australianhumanitiesreview.org/archive/issue-May-2013/teo-html (accessed 5 September 2013).
Urry, John. 2002a. "The Global Complexities of September 11th." *Theory, Culture & Society* 19, no. 4: 57–69.
Urry, John. 2002b. *The Tourist Gaze*. 2nd ed. London: Sage.
White, Mandala. 2009. "'Touring Terror': Dark Cultural Tourism in the Fiction of Khaled Hosseini." University of Leeds: unpublished paper. Cited with the author's permission.
Winterbottom, Michael, dir. 2002. *In This World*. ABC Films.
Wright, Terence. 2002. "Moving Images: The Media Representation of Refugees." *Visual Studies* 17, no. 1: 54–66.

Chapter 6

NEW MEN, OLD EUROPE
Being a Man in Balkan Travel Writing

Wendy Bracewell

Much modern Western travel writing presents eastern Europe, and especially the Balkans, as a sort of museum of masculinity: an area where men, whether revolutionaries, politicians or workers, are depicted as behaving in ways that are seen as almost exaggeratedly masculine according to the standards of the traveller. Physical toughness and violence, sexual conquest and the subordination of women, guns, strong drink and moustaches feature heavily. This is a region where men are men – and sometimes so are the women, whether 'sworn virgins' living their lives as honorary men, heroic female partisans or, in more derisive accounts, alarmingly muscular and hirsute athletes, stewardesses and waitresses. But the notion of a characteristically masculine Balkans is not limited to outsiders. It can appear in travel accounts from the region as well, ranging from Aleko Konstantinov's emblematic fictional Bulgarian traveller, Bai Ganyo Balkanski, with his boorish disregard of European norms of behaviour (Konstantinov 1966 [1895]), to more polished travel writers who nonetheless find it useful to contrast a 'Balkan' model to Western versions of manliness. The area has not invariably been gendered as male: early German writings pictured the Balkan Slavs under Ottoman rule as feminized, unwarlike and subservient; while in the nineteenth century Philhellenes and others conjured images of Greece or Bulgaria as a defenceless Christian maiden, violated by a brutal Muslim tyrant (Petkov 1997; Roessel 2002). But associations of masculinity with the Balkans are sufficiently persistent to provoke curiosity. What purposes can they serve and how can they help us to understand the Balkans' place in Europe?[1] One way of pursuing the question might be to search for the origins of such images, whether in popular culture or in literary tropes. Tracing the genealogies of patterns of perception, however, does not necessarily tell us much about their uses and meaning. Instead, I propose a more limited investigation, exploring a handful of late twentieth-century travel accounts by Englishmen and by Yugoslavs, asking why their characterizations of place and people are engendered in

---

Notes for this chapter can be found on page 106.

particular ways in specific contexts, and what functions their gendered discourses of difference serve.

Concepts of gender become entangled with accounts of travel in complex ways. Notions of masculinity or femininity play a part in positioning traveller-narrators in relation both to their implied readers and to the objects of their commentary. The sources of authority men and women can draw upon may differ; so may the myths they use to structure their travel tales (Ulysses voyages, Penelope waits at home). Gender expectations help constitute travellers' national or cultural stereotypes: the voluble effeminacy of the Frenchman; the seductive, feminine languor of southern Europe; the dishevelled violence of a masculinized Balkans. And gendered characterizations encode – and naturalize – *relationships* between peoples and places, particularly relationships of hierarchy and power. The relation of male to female has conventionally represented relations of domination : subordination in Western culture – thus the regularity with which we encounter a gendered geography that opposes a masculine, rational and active West to a feminized, passionate and passive East.

Edward Said started off this particular line of discussion, along with much else, by noting the unchanging 'feminine penetrability' of the Orient in Westerners' accounts (Said 1978: 206). Others have enlarged on the variety of gendered characterizations of East and West – less coherent and more contradictory than Said initially suggested – but have shown how the notion of gender as a relationship keeps them linked in a permanent opposition (e.g., Lowe 1991; Behdad 1994; Schick 1999). Analyses of colonial discourse have pointed out the ways an opposition between the male traveller–colonizer and feminized colony meant that women, conquered territories and non-Europeans were made to occupy the same symbolic space in the stories that defined their place in the Western imagination (Carr 1985; Kabbani 1994; Lewis 1996). What was at issue, we hear, was not just Western elite males' self-definition, but their assertion of control over all these interchangeable domestic and foreign 'Others'. These binaries have come to seem both ubiquitous and unvarying in Orientalist (and colonialist) discourse. This is so much the case that in her book *Imagining the Balkans* (1997), Maria Todorova supports her argument that Western 'Balkanism' should be differentiated from Orientalism with the claim that persistent depictions of the Balkans as masculine place the region in a distinctive relation to the West. Unlike 'the standard orientalist discourse, which resorts to metaphors of its objects of study as female, balkanist discourse is singularly male' (Todorova 1997: 15). This is because Orientalism is 'a discourse about an imputed opposition', while Balkanism is 'a discourse about an imputed ambiguity', treating differences within Europe (ibid.: 17).

But just how stable and predictable are the double mappings of gender and power? One suggestion that things might not be so simple comes from studies of women travel writers. The neatly linked binaries of male/female, dominant/subordinate, West/East and colonizer/ colonized did not easily accommodate Western women travelling and writing in the age of imperialism. The solidarities or distinctions such women invoked could cut across categories of gender, class or race in a variety of ways: women travellers might either draw parallels between their own subordinate position and that of the colonial population, or they might assert privilege as white Westerners in compensation for their oppression as women (Mills 1991; Blunt 1994; McClintock et al. 1997). Women travel writers could and did choose to write as adventurer-heroes, or as authoritative aesthetic 'beholders', in the process subverting and exposing the assumptions of discourses conventionally coded as masculine (Lawrence 1994; Bohls 1995).

Still, is it so different for men? The consistency of masculine discourses is sometimes taken for granted, as is their stability over time, in order to set up a foil to the differences that emerge from other positions. But understandings of gender, race and class shift and change, altering the meanings of the rhetorical patterns they underpin. Men differ among themselves, by age or sexuality or nation, and so can take widely differing positions with relation to the intersection of gender and power. And male travel writers, even white middle-class ones, can deliberately satirize and challenge myths of masculinity and write across the grain of dominant understandings. Different ways of being a man in travel writing, and the implications for travel writers' gendered geographies, deserve more explicit attention.

In what follows, some of these issues are explored through two sets of travel accounts: the first by Englishmen describing their adventures in the Balkans (and eastern Europe more generally) after the events of 1989; and the second by Yugoslav writers travelling in the West in the 1970s and 1980s. I do not propose to compare their conceptions of the Balkans in any detail; still less to trace possible linkages between the images that they deploy. However, the texts are comparable in that both sets of writers use depictions of gender relations, and particularly issues around masculinity, as an important element in their representations of identity and place. For both the Englishmen and the writers from Yugoslavia, ideas of manliness define and reinforce divisions between us and them, self and other, norm and deviation, domination and subordination, and West and East. At the same time, the writers link depictions of masculinity and of otherness for other purposes, suggesting personal or local interests that cannot be understood solely in terms of relations between East and West. Each set of texts raises related points about the ways gendered discourses of difference have been used in Western and Balkan travel accounts. And when they are considered together, each casts an unexpected light on the other.

Englishmen in the Balkans

The fall of the Berlin Wall was triggered by the raising of restrictions on travel to the West, but the events of 1989 were also the occasion for a less dramatic current of travel from West to East. As well as works by foreign correspondents and academic pundits bearing witness to the collapse of the socialist regimes, a host of accounts by travellers in the 'new Europe' began to appear from the early 1990s. In contrast to texts asserting the authority of long acquaintance or earnest study (e.g., Thompson 1992; Garton Ash 1993; Kaplan 1993), a cluster of books by Englishmen presented a deliberately inexpert and dilettante-ish perspective on the region.

These journeys are unconventional, if not downright eccentric – though seldom without more serious ambitions. Giles Whittell, in *Lambada Country* (1992), cycled to Istanbul, while Jason Goodwin (*On Foot to the Golden Horn*, 1994) and Nicholas Crane (*Clear Waters Rising*, 1996) headed for the same destination on foot, all in pursuit of a deeper understanding of Europe's divisions and of themselves. Rory MacLean, in *Stalin's Nose* (1992), drove a Trabant through eastern Europe accompanied by an aunt and a pig, seeking to grasp the tension between individual responsibility and collective evil in eastern Europe's past, while Tony Hawks (*Playing the Moldovans at Tennis*, 2000) demonstrated the power of a can-do attitude by pursuing a bet that he could defeat the entire Moldovan football team at tennis, with the loser of the bet to sing the Moldovan national anthem, naked, in Balham High Street. Robert Carver (*The Accursed Mountains*, 1998) went to Albania looking for 'somewhere right off the map, with no tourists or modern development', and returned to 'explain Albania to the West'.

In *Lambada Country*, Giles Whittell adopts a typically self-deprecating tone:

> It was 1990. Like thousands of others, I wanted to see Eastern Europe before it disappeared and became a mere annex of Western Europe. In particular – and this was as close as I got to what you might call a line of enquiry – I wanted to go to those parts which other forms of transport might not reach. That is, down minor roads, up steep roads, along dirt roads. Once there, the idea was to ask whether the revolutions had made a difference – to the beer, the newspapers, the prospect of going to work on a Monday morning, the way policemen spoke to you, the availability of bicycle spares [...]. Then there were the grandchildren to think of. ('Yes,

Tom, what your father says is true. Many years ago I rode a bicycle to Istanbul ...') This being a bicycle trip, there was also, for the first time in my life, the prospect of developing some real muscles. (1992: xiii–xiv)

While these writers laugh off any claims to authoritative knowledge, they offer not only entertainment but also the implicit promise that the reader will in fact gain an insight into the region that other, more conventional commentators cannot provide – precisely because of the authors' combination of amateurism (and thus a paradoxical authority derived from apparently innocent observation) and their predilection for the unbeaten track (and access to the ordinary and random, and therefore truly 'authentic'). Furthermore, they attribute an unknown quality to the region travelled to, not reducible to its postsocialist status. This lies at the heart of its attraction for these writers. As with Carver, these travellers' routes lie 'right off the map'; they see a 'chance of finding places up there'; that is, places that are not yet like the rest of Europe, places that are – in short – different. And that difference is expressed, in all these travel accounts, in terms of gender and, particularly, in terms of masculinity.

But even before these travellers address differences of place, they deploy a series of gendered differentiations to establish their own identities. The writers' sense of what it means to be a man emerges, first of all, in the ways they set themselves in contrast to traditional images of the male traveller. For most, this takes the form of the heroic adventurer and gentleman scholar of the British imperial past. Their texts are haunted by men like Patrick Leigh Fermor, who set out in 1933 from the Hook of Holland to travel to Constantinople on foot. Carver sits at Leigh Fermor's feet, seeking advice on where to escape tourism and modernity; Crane solicits his approbation for his plan to hike Europe's mountain ranges; Goodwin cites him as an inspiration. Whittell mentions him only in passing in his own journey to Constantinople, but his account in many ways can be read as a comic inversion of the older traveller's journal: Whittell paints himself as inept where Leigh Fermor was omni-competent, pursued by the Lambada rather than accompanied by folksong, tumbled by orange-fingernailed tarts in campsite *cabanas* rather than tumbling peasant girls in haystacks, and passed from one Hungarian household to the next – for arguments over apartment kitchen tables rather than repartee in aristocratic salons. But similar elements of parody and bathos also structure the other accounts. These travellers recall the heroic men who preceded them only to present themselves in comic contrast, as incompetent, clownish antiheroes, who have no idea what to pack (Goodwin's mountain of hiking equipment includes both silk and thermal underwear and five bars of Bendick's Sporting and Military chocolate, which he contrasts to the rucksack holding

Wendy Bracewell

'a toothbrush, an apple and a pair of socks' carried by 'our predecessors' (Goodwin 1994: 13); whose embarrassing physical incapacities are detailed (Whittell's cyclist's crotch-rot, Goodwin's diarrhoea, Crane's endless list of excruciating ailments); and who cannot do anything without help.

Yet, for all their self-ridicule in comparison to the heroes of the past, these writers make sure to present themselves as travellers – never as tourists. They may not live up to the standards of the men who built the Empire, but they follow the same tradition of enterprise and adventure, although somewhat diffidently, perhaps, when it comes to the notion of an Empire, whose passing they note by blaming their 'khaki empire garb' for incubating sweat rash (Whittell 1992: 104); by poking fun at the anachronistic language and pretensions of know-nothing British diplomats, summed up by the figure of 'Carruthers, Our Man in Tirana' who is reduced to an embassy of 'three small rented rooms in an office building' (Carver 1998: 151–54); and by lampooning an annoying sidekick's resemblance to all the 'maverick Foreign Office Arabists known through the bazaars from Alexandra [sic] to Lucknow simply as The Englishman' (Goodwin 1994: 8). These writers are more confident about being adventurer-travellers when it comes to comparisons with their own contemporaries, particularly in chance encounters with Western tourists, who are usually likeable and well intentioned, but basically consumers of guidebook experiences, whether passive and uncomprehending or earnest and overprepared. Domestic social distinctions are not foregrounded in this familiar traveller/tourist dichotomy – but the travellers' dogged egalitarianism is not allowed to obscure their own very evident social advantages. (Carver does highlight his own superior daring as a traveller by dismissing the travels of 'waffling old Etonians on bicycles' (Carver 1998: 331), but he also ensures we know about his own public-school background.) National differences are a more acceptable substitute for social difference. Germans and Americans are particularly apt as foils to our heroes: German fellow-tourers are equipped with 'Lennon specs, Goretex overgarments, hyper-rugged bikes loaded for total self-sufficiency' (Whittell 1992: 172), or at least 'a long, hard sausage' (Crane 1996: 85), while American volunteers and tourists are unremittingly groomed and hail-fellow-well-met, as well as loaded with camcorders, super-soft sneakers, and money (Hawks 2000: 43, 144–45; Carver 1998: 276–77; Whittell 1992: 181). But what are all these advantages compared to their own gentlemanly English virtues of amateurishness, stoicism and whimsy? For all their self-deprecation, the young Englishmen are nostalgic for a vanished imperial masculinity that would give purpose and legitimacy to their anachronistic attempts at adventure. Crane sums it up for all of them, comparing his experiences to those of Leigh Fermor: 'I was too late' (Crane 1996: 328).[2]

A certain nostalgia is perceptible, too, in the way these writers establish their relations with Western women. Several of the accounts record the shadows of ex-girlfriends and wives left behind, who have no intention of waiting for the traveller to return and are more interested in getting on with their lives than in taking part in boy's games. Goodwin, unusually, is accompanied by his girlfriend and future wife: she thought his plan of 'bridging the gap between traveller and indigene' by 'grubbing about in mud and boots' was 'dim', but realized she would have to come along if only to ward off possible dangers (Goodwin 1994: 10). Other Western women met on the road are self-confident travellers, more frightening than attractive. Their emancipation seems to imply a corresponding emasculation on their menfolk's part. Carver is quite explicit about the equation:

> Over the last thirty years the gradual feminization of society in Britain, and most of the formerly macho northern European democracies such as Holland and Germany, had blanded men down to an acceptably low-testosterone product, suitable only for occasional use by the quasi-liberated women, as and when required. (Carver 1998: 184)

He goes on to contrast 'use-and-chuck, Kleenex-style, Euro-wimps' to 'real, old-fashioned Western males, authentic gas-guzzling pre-feminist models' – whom he sees as now largely extinct (Carver 1998: 184). Carver is particularly scathing, but anxieties about what it means to be a modern man in a society where women are understood as equals are shared by all these writers.

The part played by local women is superficially similar to that of Western women. The travellers are constantly being taken in hand by local women who rescue them and solve their many problems. On the one hand, their calm assurance, like the confidence of the Western women, gives substance to the travellers' claims to incompetence. But while Western women do not need the travellers, these women do. Skilled, dynamic and capable Balkan women are regularly described as deserving better than the limited opportunities available to them – and as needing to be rescued. These are not instances of Gayatri Spivak's classic colonial romance: 'a white man is rescuing a brown woman from a brown man' and in the process justifying colonial rule as the protection of victimized women (Spivak 1988: 297). These abortive romances never reach fruition: an Englishman is *failing* to rescue a Balkan woman. But the reason is always the same: the woman's material motives and the fear that the man represents nothing but 'an escape route from this country' (Carver 1998: 184; Hawks 2000: 159). These Englishmen hesitate at the thought that these relationships might depend on something other than their own individual qualities: they want to be loved for themselves alone, not because they might stand for Western political and economic might. This is a masculinity that still romanticizes masculine power and feminine dependence (placed in piquant contrast to the independence of the modern Western woman). Yet, also in line with older understandings of bourgeois respectability, the travellers insist on hiding any hint of a more calculated transaction behind an ideal of sentimental reciprocity.

Encounters with prostitutes follow the same principle, but are much more straightforward. Not only does the sentimental Englishman never pay for sex, the very prospect of such a transaction confuses his reactions entirely. Nicholas Crane is importuned by an attractive Bulgarian prostitute : 'You, me – sex!' (Crane 1996: 357–58). He feels that a woman is a fair reward for his travails: 'Didn't I deserve a dose of delicious coddling?' But all he can think, as things get out of hand, is: 'What about the transaction? Do you arrange a price before, or after? Would she take a traveller's cheque or demand dollars? Is it a flat fee or do they charge by the minute?' He comes to himself by remembering who he is: 'No, thank you. I'm English'. The woman is incredulous: 'English? Polish? What the difference?' She does not understand. For Crane, men are *not* all the same, and for him, as for other postimperial travellers, it is crucial that the Englishman maintains that difference by preserving the boundary between eros and commerce.

Crane makes the prostitute and his encounter with her stand for Bulgaria in general: she smells of roses, the symbol of Bulgaria, 'not a subtle hint of roses, but an over-powering pall which must have been applied as body-lacquer with a high-pressure hose', while her cleavage echoes the countryside itself: 'wasn't there a Valley of the Roses in the Stara Planina?' The abortive romances between local women and English travellers regularly serve as a metaphor for the encounter between the East and West: these travellers are obscurely disturbed when it turns into a commercial exchange. Expecting gratitude and even love as the reward for deliverance, they find that the East has more material interests and knows what she has to bargain with. It is not the relationship that their nostalgia seemed to promise. The new Englishman can find the new Europa rapaciously capitalist, even if deliciously feminine.

These versions of modern, postimperial English masculinity are thrown into higher relief by encounters with local men. The travellers are constantly comparing themselves to their hosts and, once again, failing to measure up according to standard virility indicators. A whole series of set pieces sees them outmatched in competitions over drinking to excess, shooting guns off recklessly, driving dangerously, pursuing women, sword fighting or moustache growing. These encounters define a stereotypical machismo, compulsively competitive and rooted in physical or sexual prowess. The English travellers treat this hypermasculinity as both familiar and exotic. It is an aspect of eastern Europe's backwardness, a marker of a phase that their own society has passed through, as corny and outdated as the Lambada; but at the same time different, engrossing, simultaneously repellent and attractive. These may be 'real' men, but theirs is a version of masculinity that the writers see as lagging behind their own society's gender norms. Such machismo makes the region somehow less modern and less 'European' – though local men are never compared to the Spanish or Italians, nor indeed to working-class British men.

The reaction ranges from fear, through a sort of aesthetic appreciation, to attempts at emulation – sometimes all in the same text – as the writers position themselves and their hosts on a spectrum of manliness. Carver, for example, weighs up the risks of 'failing a local test of machismo' when offered a potshot with the bus driver's automatic on an unscheduled Albanian rest stop; he turns his back and walks away from the gaggle of armed Gheg passengers, trusting the camera-bulge under his shirt to suggest his gun-carrying credentials, simultaneously terrified and proud of his grasp of the rituals of Albanian manhood (he reasons that they will not shoot him if there is a chance of hitting the boy standing in his path and thus causing a blood feud) (Carver 1996: 229, 233–36). For Carver, it is a violent, patriarchal and irresponsible masculinity that defines Albania (the 'Land of the Eagle' is summed up in terms of its 'Sons': 'hospitable rapists and elegant torturers, welcoming robbers and wife-beating family men' and so on (Carver 1998: 337)). Other encounters are less melodramatic but equally emblematic: Rory Maclean falls in with a local man, Kristan, in Romania, and finds himself being instructed in womanizing, ingenious methods of counterfeiting cigarette packets, and black-market transactions in a chapter entitled, ambiguously, 'Riding with the Best Man' (ostensibly referring to conductors who take bribes on the railway, but also summing up Maclean's experiences with Kristan) (Maclean 1992: 169–78). Whittell is floored by *slivova* in Bulgaria while his smuggler companion drinks until morning and stumbles in to bed, but still manages, 'incredibly, to swivel back to the door and open it before urinating and vomiting' (Whittell 1992: 185–86). Whatever the sphere of action, the Englishmen usually come off second best by local standards of manliness.

However, there are limits to self-deprecation. Their hosts may possess an old-fashioned machismo, but this only highlights the qualities that the Englishmen see as their own defining features. The first of these, touched on by most of the writers, has to do with money and work. While local men may be virile in physical and sexual terms, they are emasculated by being poor – especially when they cannot provide adequately for their families. The travellers are constantly irked by the assumption that they, in contrast, are rich. Several contrive to be perpetually short of cash as part of their adventures on the road, lessening the apparent distance between themselves and their hosts and allowing them to accept hospitality gratefully or to reward it at their own whim – but without noting that their temporary penury is entirely self-inflicted and conceals their ability to spend their time at their leisure. (None of these writers tells us how his journey is financed.) They hesitate to interpret economic power in itself as evidence of a fundamental difference between themselves and the locals. They attribute the poverty they see to the collapse of the economy and the welfare network resulting from large-scale political and economic change: the structured inequalities between East and West cannot easily be translated into evidence of personal qualities or codes of behaviour. But this analysis is constantly undermined by the way they comment on the local men's passive, fatalistic acceptance of their circumstances, or else their desire for easy money – for something for nothing. It is not money that differentiates these Westerners from their hosts, but the work ethic. Even Tony Hawks, who admits that his aim of beating Moldovan footballers at tennis is a frivolous waste of time and money, in dubious taste in such a poverty-stricken country, differentiates himself from his Moldovan acquaintances on the basis of his conviction that any difficulty can be overcome by effort. His greatest victory in Moldova is not trouncing the footballers, but getting the teenage son of his host family to laugh at his antics (Hawks 2000: 139–40) and finally, as his crowning achievement, to admit the power of his positive philosophy and to take Hawks as his model of manhood for the future (ibid.: 249). The notion that passivity or inertia is a characteristic of local men is underlined by the pointed contrast made with enterprising (if frighteningly rapacious) local women intent on achieving change for themselves and their own families by any means possible.

This suggests a second quality that differentiates the Englishmen. The travellers notice the ways in which local women are exploited by their menfolk and by society as a whole. For Whittell, the typical rural family 'seems to consist of a hospitable alcoholic husband and a haggard, sober, overworked wife. In taking advantage of the hospitality I am abusing the wife' (Whittell 1992: 93). These Englishmen see the inequalities that go along with a division of gender roles and deplore them, open-mouthed at women who are proud of being dominated by their husbands (Goodwin 1994: 153). They accept their ministrations self-consciously, not as a right but as an embarrassing throwback to a 'colonial childhood' (Carver 1998: 92), only allowing their 'impeccable credentials as a politically correct male' to slip briefly when pushed beyond endurance by a woman whose ability to annoy outweighs the evidence of her oppression (Crane 1996: 223). Local men are presented as taking the patriarchal gender regime for granted. The Englishmen can see themselves as feminists by contrast. Straightforward and unreflexive, local men provide a contrast to the Englishman's stance of self-awareness and self-doubt. Balkan men may be shown as struggling to adjust to economic and political transition, but they have not noticed that it is no longer so simple for modern men in other spheres either.

These travel accounts should be read in the context of the so-called 'crisis in masculinity' in 1980s and 1990s Britain, a state of affairs usually attributed to feminism, changes in the economy (the decline in traditionally male-dominated industrial sectors, the growing presence of women in the labour market) and an increased acceptance of alternative sexualities. Responses ranged from the long-standing feminist critique of patriarchy to popular attempts to define a 'new man' (Connell 1995). The travel writers under discussion, too, use their adventures abroad to play with a variety of notions of manliness and stake out a revised version for themselves: more enterprising than their stay-at-home peers; more daring than the Western tourist with his package holiday; nostalgic for the privileges and certainties of an imperial past, but at the same time more responsible, more emotionally literate, more feminist, more politically correct than the standard Balkan male. The eastern Europe depicted in these accounts serves largely as a backdrop, painted in such a way as to foreground the revised English male identities being developed through travels in the East. While being a 'Euro-wimp' might be a source of anxiety to the Englishman abroad, the new masculinity he is in the process of mapping out is at least superior to the outdated and superseded models he encounters on his travels.

The character of the depiction does not match the mappings of gender and power conventionally attributed to Orientalist or colonial discourses: Western supremacy is not asserted in terms of the general pattern of 'the demasculinization of colonized men and the hypermasculinity of European males' traced elsewhere (Stoler 1991: 56). But just as studies of Western women travellers have drawn attention to the diversity of their purposes and circumstances, so too an examination of these male travel writers shows a variety of factors at work. Looking at the images used by Englishmen in the context of the 1990s 'crisis in masculinity' helps show why the usual gender polarities of alteritist discourse might have been reversed in this way, at this particular juncture, and using this particular tone. Parody and inversion of conventional expectations, motivated by changes to Western middle-class gender norms, underpin the discourses of these travel accounts.

It is sometimes precisely this 'revised' masculinity that works the hardest to sustain a hierarchical relationship between East and West. This is not always overt: straightforward assertions of superiority have become suspect, in geopolitics as in gender and class. The writers' self-parody as hapless anti-heroes, undercutting and deflating their own pretensions, serves as a self-defensive strategy. But their self-deprecating depictions still have consequences:

their stories draw on and reinforce older notions of geocultural difference, and they evaluate their simplifications and generalizations in moral and hierarchical terms. The West is still the superior norm and eastern Europe and, even more, the Balkans represent the inferior deviation. These English travellers may be 'new men' in contrast to older models, but what they give us is definitely an old Europe.

Yugoslavs in the West

The line of analysis followed above fits comfortably within the Orientalist (and Balkanist) paradigm (Wolff 1994; Todorova 1997; Goldsworthy 1998). However, scholars have insisted on the distinctiveness of the relationship between the Balkans and Western power: the absence of direct colonial rule in particular has meant a corresponding stress on Western *cognitive* hegemony in the region (specifically in the Gramscian sense of hegemony as the consent of the dominated). Claims have even been advanced that the categories of self-identity have been colonized by Western modes of thought. Whereas beyond the borders of Europe, 'the logic of domination is imposed by colonial rule', in the Balkans it seems to be 'the immanent logic of self-constitution itself that generates the incapacity to conceive of oneself in other terms than from the point of view of the dominating other' (Močnik 2002: 95). Internalization of Western ideas of the Balkans, and the notion of being *in* Europe but not wholly *of* it, are blamed for inflicting a whole series of traumas associated with ambiguity, assessed in terms that range from 'self-colonization' (Kiossev 2002) to 'self-stigmatization' or 'geocultural bovarism' (Antohi 2002). In such analyses of the Balkan variant of Orientalism, the idea of 'the Balkans' becomes essentialized even as it is deconstructed: understood in abstract structural terms, as a dark destiny imposed on southeast Europeans by the inescapable logic of centre and periphery.

Such notions can appear in a slightly different light if we examine depictions of masculinity, 'Europe' and 'ourselves' in different contexts. It is to this end that a second set of travel accounts, published in Yugoslavia in the 1970s and 1980s, will now be explored. At that time, travel accounts of western Europe and the United States were a well established genre in Yugoslav publishing, accorded a degree of literary prestige and often published by authors with a reputation in other genres. Examined below are three such accounts, by a fairly cohesive group of established writers. Momo Kapor is a Serbian novelist and journalist; his travel account, *Skitam i pričam* ('I Wander and I Talk') (1979), consists of brief, anecdotal sketches, grouped by region (Dalmatia, Europe, the USA and Belgrade), which use his travel experiences to pass judgement on cultural difference. Moma Dimic, a poet and novelist from Serbia, had also previously published travel accounts; his *Monah čeka svoju smrt* ('The Monk Awaits his Death') (1983) recounts his travels in Greece, western Europe and the USA, mixing literary encounters with cultural critique. Ivan Kušan, a Croatian playwright, novelist and children's writer, presented travels through western Europe, Russia and the USA in *Prerušeni prosjak* ('Beggar in Disguise', 1986) as erotic picaresque (a genre that the cover blurb identifies as 'globetrotterotica').

Accounts of the West had their own legitimacy in post-Second World War Yugoslav literature: they were frequently set in parallel to travels in the socialist bloc, with the point being to highlight ideological contrasts between Western capitalism, Warsaw Pact socialism and Yugoslavia's own brand of socialism. However, such distinctions became less emphatic from the mid 1960s, with the easing of ideological strife and the spread of international détente. In these accounts, differences are defined less in political-ideological terms than in civilizational ones. The writers draw on and reshape ideas of East and West, the Balkans and Europe, and Europe and its US other. Yugoslavia is the primary framework for the authors' self-identification in these texts, with greater or lesser emphasis on a Serb or Croat national affiliation. But the authors regularly blur any more specific definition of identity (whether Serb/Croat, Yugoslav or Balkan) by the frequent use of 'us', 'ours' or *po naški* ('the way we do it'). Who this 'we' is needs to be deduced in each context from the defining others – Europeans, Americans or Westerners for instance, or Russians and socialist fraternal 'brothers', or other Yugoslav and southeast European nations – all located in gendered terms. Balkanness, masculinity and difference are linked in each of these texts, as analyses of Balkanist discourses might lead us to expect. But at the same time, the Yugoslav writers actively *use* these equations to make a variety of claims that are difficult to understand solely in terms of self-stigmatization or imaginative colonization.

In these Yugoslav texts, sex is used extensively as a topos for representing difference and fleshing out conceptions of 'us' and 'ours'. Especially for Kapor and Kušan, contact with the West (and the rest of the world) is presented primarily through erotic encounters between 'our' men and 'foreign' women. Western women pose a challenge to the traveller's virility and authority, as well as a means to cultural mastery. 'Having a woman' is, predictably, a way of becoming a part of an otherwise unattainable world. Kušan's Parisian Bernadette is thus different from his Yugoslav *petites amies*; he needs her (and Monique S. and Jeanne and Colette and so on) in order 'not to feel a tourist'. 'I had to have a native Bernadette like this one to give me the illusion that I had at least some lasting root connecting me to this quay, to which I had no right at all [...]. I'm no longer completely a foreigner here' (Kušan 1986: 93).

But sexual success is presented not in terms of masculine 'possession', but rather as passing an examination. The West is thus feminized and eroticized, but it is by no means subordinate or inferior: these women are shown as independent, choosy and critical. They may appreciate the travellers' Balkan virility but they insist on their own standards in other spheres such as gastronomy, hygiene and fidelity.

Western men scarcely appear in these encounters. At most they appear as vague collectives, the generalized voices of Kapor's stereotyped conversations, or distantly observed 'solid citizens' or 'queers'. They are not even rivals. If the Englishman's failed Balkan romance can be compared to Spivak's colonial plot, the Yugoslav romances are not comparable to Frantz Fanon's reverse fantasy, in which the possession of white (read: Western) women by black (read: Balkan) men constitutes both revenge against white men and the appropriation of their civilization and dignity (Fanon 1967). Here, women are not just the terrain upon which an East/West struggle is enacted; they have their own interests. And Western men are already emasculated in advance. The same is not true of Third World men, or men of other Balkan nationalities, who regularly appear in these texts as potential rivals. Their presence further locates the sphere of difference the writers are constructing. Kušan, for instance, finds the most threatening masculine challenges to his mastery and self-esteem in Paris in the person of a well-endowed Moroccan who also has a big wallet and native French (shared non-alignment does not prevent them aligning in their rivalry over the Englishwoman Gill) (Kušan 1986: 100–102). In America there is Petru, a tall, dark and handsome Romanian with British-accented English as well as near-native French ('as though Romanian were anything more than French à la Dracula, a Latin language on the lips of Slav bats' – ibid.: 209). Russians ('Scythians') do not threaten so much as thwart: Kušan's whole Russian journey is a tale of incapacity, which gives substance to references to Soviets/Scythians elsewhere in his text.

Much more striking than encounters with Western men, and more fully developed, are the writers' relations with 'our' own men, whether travellers, émigrés, friends or companions. These men function as yardsticks against which our heroes measure their own European acculturation (or the ability to 'pass') and maleness. Kapor's 'Piter' and 'Džordž', with their successes with women and in business and, equally, their distinctive capacity for pleasure, disrespect for bourgeois behaviour and their nostalgia for home, seem to serve as surrogate selves, placed in the narrative to show an idealized version of Balkan manliness abroad in contrast to denationalized émigrés with their foreign wives and children. Kušan has a whole series of sidekicks: most exemplify the shamelessly inassimilable Balkan man who takes his culture with him wherever he goes, picking up European habits (swapping his *Drava* cigarettes for *Gauloises*) but never adjusting his own values or abandoning his capacity for gross physical pleasure; there to show just how far Kušan, in contrast, has been changed in his encounter with the West.

'Our' women also play a prominent role in these travel tales, not just as sexual partners, but as emblems – for good or ill – of the differences between home and abroad. They are markers of all that the traveller has left behind (or wants to shake off). Kušan's Vera is his companion on his earliest French travels but is outgrown and discarded (along with a taste for home cooking, and the need to shop for textiles to take back as gifts). Later, Branka's sexual attractions wax and wane in Kušan's eyes in inverse proportion to the availability of her German or American rivals, registered by approving or derisive assessments of her 'Balkan' qualities – her mentality, appetite and bottom (sometimes 'a peasant girl's [...] firm and compact', but at other times described as 'like the bald, fat, red cheeks of our village alcoholics' [Kušan 1986: 166, 180]). Kapor's Snežana has left the Balkans behind: after wasting three years with a drunken charmer at home, she has sought security with a thoroughly Helveticized anaesthesiologist ('everyone thinks he is a native Swiss, he's so punctual; the greatest compliment that a barbarian from this unhappy part of Europe can receive'). But how can she be happy? 'Don't you want to quarrel like a human being, to break all the dishes, to sing while you wash the windows, to borrow coffee or oil from your neighbours, to eat watermelon without a knife or fork, and wash your fingers in the river?' (Kapor 1979: 123–24).

These Yugoslav authors, like the Englishmen discussed earlier, draw gendered maps of the Balkans and the West, with contours that are sometimes strikingly similar. But while those Englishmen use the Balkans as a point from which to reconsider modern masculinities, these Yugoslav writers are more interested in using the mantras of masculinity in order to comment on 'us' and 'them' – on the meanings of the Balkans, Yugoslavia and Europe. While they agree on the fact of difference, and even on some of its markers, they do not all come to the same conclusions. The Yugoslav writers produce a range of assessments of East/West differences and use these concepts to place themselves in different ways, for a variety of purposes.

Comparing accounts of prostitutes and sex shops, a recurring topic in these travel accounts, provides a convenient way of illustrating this point. A scandalized description of prostitution had been a standard set-piece in early socialist accounts of the West, serving to condemn the degradation of women under capitalism. By the 1970s Momo Kapor, in contrast, can poke fun at prudish socialist travel accounts from the 1950s describing a 'Parisian hell' where, 'just imagine, the girls in the Pigalle sell love for money!' Now the world is no longer so simply divided; 'the West hasn't been sunk in eternal darkness for a long time, and the sun doesn't invariably rise in the East'. Instead, Kapor identifies another pattern of exchange, with 'our' young men offering Paris 'their fresh Balkan blood and new ideas that they never dream have long since been being taught here in the primary schools', writing poetry and living off susceptible French women of a certain age who translate their verses. 'Older women in Paris are in luck, as long as our boys go there! Any one of them can find an Edith Piaf who remembers better days. And one day his French verses will be retranslated into Serbian and our grandsons will have to memorize them in school'. This allegorical relationship between French women and Balkan men turns on an established geocultural difference: 'our unrequited love for Paris is as old as our provincial yearning for world-wide fame' (Kapor 1979: 114–16).

Kapor hints at the way he sees the relationship between the Balkans and the rest of the world in his European travels, but he develops his vision of 'us' and 'them' most fully in the United States section of his account, entitled 'The Marquis de SAD' – SAD being the Serbo-Croat abbreviation for the USA. Here again, selling sex becomes emblematic. Kapor spends an entire day on New York's Broadway and 42nd Street, escaping 'the crowds of transvestites, homosexuals and whores', as well as a cold north wind, by visiting a peep-show. Though his description echoes the disapproving sociological investigations of earlier accounts, he also makes unfavourable comparisons with sexier European shows: 'Not a trace of the coquetry of the Parisian ladies in the Pigalle, none of the fleshy femininity of the Antwerp prostitutes in their red window displays ... Nothing apart from the deadly boring parting and closing of legs, twisting of bodies and monotonous changes of position, like an anatomy lesson'. But the real point of his observations is elsewhere: 'I'm almost more interested by the faces belonging to the feverishly burning eyes peering through the little openings than I am by the tired movements of the two enslaved female bodies offering themselves listlessly to this painful curiosity' (Kapor 1979: 164–65). What sort of men are these Americans, to satisfy themselves in such a manner? The peep-show sets up a dichotomy between US perversion and *European* pleasure that persists throughout this section, with work- and money-obsessed US men contrasted, for example, to 'a whole naïve army of European lovers, who are still up for love. They haven't forgotten how to be tender, coarse when necessary; rogues hungry for love, just like their old continent, unpredictable and crazy. They aren't ashamed to buy yellow roses at six dollars a stem and to whisper tender words' (Kapor 1979: 194). They are doomed to disappointment, though: the US women only have time for this sort of thing at weekends. And there is always the danger of losing your true self in this alien land. Kapor lunches with 'one of our countrymen' in San Francisco, but he had become 'so refined and delicate that he had gone completely vegetarian. His great-grandfather had hoisted a live Turk in his teeth, his grandfather had eaten roast ox for dinner, his father had snacked on a half a lamb from the spit, his mother had raised him on fresh liver, but he says: "A vegetable cocktail, please, with extra carrots"!' (Kapor makes a point of ordering the beefsteak, very rare: *he* is not going to be a traitor to his sex or to his origins) (Kapor 1979: 158–59).

Kapor's scale of values sets appetite, abandon and the capacity for pleasure against reserve, control and an obsession with time and money, in what might be called an 'affirmative Balkanism'. But, strikingly, it's not always Balkan. In Kapor's US travels, 'ours' is equivalent to 'European'. Sometimes this is ironic: 'The Americans have a "European complex". And it's well known that we are part of Europe. That means they have a complex about us, too. Oh, what sweet consolation!' (Kapor 1979: 175). But elsewhere the distinction is between a 'European' sensibility and 'ours'. In his descriptions of Paris, in his 'Sentimental Journey' through southern and northern Europe and, even more, in his observations of domestic and foreign tourists in Dalmatia ('Summer') and in his celebrations of Belgrade, 'ours' is interpreted as Balkan, Yugoslav or Serbian, depending on the context. The pair of compasses measuring this symbolic geography have their pivot planted at the centre of a circle that can expand or contract, depending on the alliances or exclusions that are implied. But virtue lies at the centre for Kapor: 'Balkan' is beautiful. He sums up his position in considering the concepts of East and West in his 'Sentimental Journey': 'Since we happen to live in between the East and West, we believe that truth and the measure of man lie somewhere in the middle' (Kapor 1979: 92).

While Kapor is interested in evaluating the divisions between home and the world, Moma Dimic's preoccupations are more local. His account of Hamburg dwells on the Sankt Pauli district, but without much trace of moral condemnation. Prostitution is primarily an economic activity, one that has 'done its part in Germany's postwar "economic miracle"' by soliciting the financial contributions of 'our Gastarbeiters', among others (Dimic 1983: 117). The description of the possibilities on offer and the conventions governing the transaction is completed – and given animation – by a quizzical vignette of a pair of compatriots:

> Once, in nearly the same place [the Eros-Centre brothel], I saw two of my countrymen. The farther north you go, the easier it is to pick out our men. Lean, no longer quite so young, they wander indecisively through the voluptuous twilight, gape, and stare endlessly at the girls on display: the heart shapes formed by their unclad buttocks, the tender skin of their thighs, their uncovered shoulders. Anything above and beyond that would cost too many of their Gastarbeiters' marks – carefully hoarded but never enough. They will go round all the courtyards and the streets with girls on

display several times. They will approach the doors of topless restaurants and variety shows featuring female mud-wrestling or boxing, timidly, but these too will be too expensive for them. They begin their free Saturday night with such excitement and such luxuriant nakedness, but they end it alone, in a cheap Oriental café, with a piece of *burek*. (Dimic 1983: 119)

Dimic's countrymen have little in common with him besides nationality. He takes for granted what they desire but cannot afford, whether sex or other Western consumer goods. These emasculated Gastarbeiters are doomed to remain mired in the Balkans wherever they may actually travel, work or live, subsisting on *burek*, that emblematic Balkan fast-food pastry. Dimic, with his experience, savoir faire and economic power, has access to other ways of life. In such texts, Dimic – like other writers – uses concepts of East and West to mark out (and perpetuate) the social distinctions that existed at home, as well as abroad, between an educated elite and a working class. Yugoslavia's Gastarbeiters had their own extensive experience of the West and made this evident at home (not least in the form of hard-currency savings accounts). But experience of the world counts for nothing in Dimic's account unless it translates into discernment and the power to choose. In spite of his use of the word 'ours', Dimic places himself on the other side of a cultural divide from his 'oriental' compatriots. He is aligned against the Balkans, alongside the cosmopolitan men of the world who know how to enjoy Hamburg's opportunities and have the means to do so, rather than with his working-class compatriots.

Ivan Kušan is much more ambivalent about being Balkan. Neither his attempts to master the Western world through his sexual adventures with its women nor his attempts to pass as a Westerner are presented as successful; each failure contributes to his sense of inferiority and lack of entitlement to the life he tastes in the West. Hence the title, with its inverted reference to Odysseus:

I put on a pretence that I wasn't only here [in the West] by chance, but had been here from the beginning. The point of my mimicry wasn't to hide, but rather not to stick out as an undesirable, inassimilable intruder. Not disguised as a beggar, so that the suitors (the swine) wouldn't guess, but disguised as a man of means, a beggar in disguise. (Kušan 1986: 363)

Each failure serves to show where he really belongs: 'I had turned out to be what I really was: a little Balkan scribbler megalomaniacally trying not to fart, in the face of the world's iron indifference' (ibid.: 108). Kušan can never be a Western man of the world. He is constantly dragged back by cultural and geopolitical circumstance, his self-doubt reinforced by Yugoslav socialism's unattainable promises of a paradise of 'tempting fruit, hams, bottles and, above all, beautiful naked girls' (ibid.: 104) and the West's abandonment of his half of Europe 'to the favour and disfavour of the Scythians' (ibid.: 107). But neither is he the same as his Balkan compatriots who do not even worry about such differences and enjoy themselves without anxiety. He makes this clear in a text that, once again, focuses on prostitution. He cannot emulate a colleague who feels completely 'at ease' in Western brothels ('he didn't even bother to think in our language').

> Standing in front of those famous Amsterdam windows, I recalled how, when his eye fell on some modest 'housewife' in her display window, darning and reading her Bible on her immaculately clean bed, with gleaming sanitary ware in the background, he burst into the idyll without a second thought. The curtains closed, the light went out, the Bible thumped onto the nightstand – and my colleague ordered the complete programme, since it didn't have to be paid for in advance. Only afterwards did it occur to him that he was only carrying dinars. Since the girl's madam had never seen anything like them, she phoned the bank – and she clearly heard bad news. They carted my colleague off to the police, questioned him, expressed their disgust, and let him go. The whole time he was completely at ease, grinning childishly. He was still enjoying it the fifth time he told me this story, behind the bar of a luxury Frankfurt brothel, while we sipped *Sekt* from real champagne flutes, naked, with our numbers on a chain around our necks (the Germans love order, they immediately take your name and give you a number) and selected our partners (I was only careful that she shouldn't be one of 'ours'). I paid in advance, naturally, but that's why he's the one who is a business operator on an international scale and is contributing – in his own Amsterdam way – to the exchange value of the dinar. (Kušan 1986: 54–55)

It is not so much economic power as it is shamelessness and the inability to see the difference between 'them' and 'us' that separates his colleague from Kušan. But Kušan, like Dimic, asserts that he is different: he has learnt to think in terms of shame and self-consciousness (he has scruples about 'our girls'), and has acquired a Westernized sensibility that distinguishes him from his countrymen, even though his sense of inferiority prevents him from asserting equality with European 'men of means'.

All three of these writers used their accounts of sex, masculinity and their travel adventures in general to elaborate a slightly rebellious and unconventional authorial persona. Their positions make more or less political points. Kapor's raffish Bohemianism was deployed against 'bourgeois values' (though his rebelliousness had limits: his critique of Western capitalist gender norms fit comfortably with Yugoslav socialist ideals, though with successive editions of his book the balance would shift to a celebration of values named as Serbian). Dimic and Kušan stood more directly at odds with the existing system in their matter-of-fact acceptance of prostitution and their frank appreciation of capitalist consumerism (and their implied criticism of the Yugoslav failure to match Western standards). Kušan, in particular, criticized Yugoslavia's in-betweenness, 'non-aligned' alongside the Third World but not quite free of 'Scythian' socialist fraternalism; just capitalist enough to introduce time-clocks to control the workers but not capitalist enough to care about satisfying consumer desire. This rebellious individuality was marketable in the 1980s, as the Yugoslav political and social system began to lose its legitimacy and new forms of criticism and dissent were finding not just a voice but also an avid reading public.

But for all their subversiveness, these writers lay claims to a masculinity recognizable in conventional terms – independent, experienced and virile. Their air of mastery asserts a manliness not always available or conceded to the intellectual in a society where the man of letters is not necessarily quite a man (v. Džadžic 1987: 180–202). Moreover, these accounts promote the intellectual as having enviable masculine advantages: the resources to travel, the experience and the taste to participate in the good life of the European elite, and aspirations to equality with Westerners. This, in turn, reinforced domestic divisions: the gap created in the text between the educated man of the world and the emasculated and orientalized Gastarbeiter legitimated the Yugoslav intellectual's claims to prestige and authority at home as well as abroad.

These discourses of difference, derived from depictions of masculinity (and from other categories), trace a symbolic geography that divides up the world in familiar ways. What these divisions are named, and how they are evaluated, however, varies. It is significant that the term 'Balkan' does not appear consistently in this second set of texts, and the writers' use of 'us' and 'ours' varies too. Who 'we' might be expands and contracts, ranging from Kapor's Belgrade or Kušan's Zagreb suburb, to a shared Yugoslav sense of belonging or a larger Balkan identity, to all of Europe, in contrast to the USA. (The English travel writers' maps, in contrast, locate 'them' on a sliding scale that differentiates the Balkans from the rest of eastern Europe only as a matter of degree.) Neither is the stigma attached to 'Balkanness' constant: Kapor's celebration of a positive Balkan Orientalism contrasts with Kušan's ambivalence and Dimic's disassociation from the label. 'The Balkans' would appear much more regularly – and in a much more consistently negative light – in accounts written in the 1990s by travellers from the former Yugoslavia, but by then the context had changed considerably. The ideas of masculinity helping to define the boundaries being erected between 'us' and 'them' need to be understood in context, too. Exaggerated masculine egos and physical appetites paired with a lack of shame or constraint might recur in representations of the Balkans from inside and from without (see also, for example, travel accounts of Greece by Henry Miller and Patricia Storace [Miller 1958; Storace 1996]), but this is neither the only available local model for manliness, nor is it limited to the Balkans.[3]

What *is* constant in both the Yugoslav and English accounts discussed there is the 'technology of place': the way that the shifting categories of us and them, the Balkans and Europe, are defined in opposition to one another, insisting on the fact of difference regardless of the content. These differences often have much less to do with great geopolitical dichotomies than with specific local divisions and agendas: 'Europe' and 'the Balkans' become weapons to be used in contests that lie much closer to home. Others have analysed the ways Orientalist stereotypes have been used in the Balkans in a process of 'nesting orientalism', intended to consign neighbouring nations to Eastern darkness while advancing one's own claims to European legitimacy (Bakic-Hayden 1995; Kiossev 2002). But representations of East and West can also trace lines of division *within* a society. Yugoslav political and ideological changes, as well as its persistent social tensions, can be seen reflected in the ways the writers in the second group chose to align themselves through their evaluations of Europe and of masculinity. The politics of gender and the unravelling of class privilege in a postimperial Britain also shaped the ways the English travel writers engendered difference.

Thinking about the choices made by these travel writers and the uses to which they put notions of East and West, us and them, and machismo and emasculation, opens up new perspectives on 'the Balkans'. It begins to seem less a matter of a Western projection imposed upon the region or a traumatic geocultural destiny (however such claims might suit Kušan's defensive self-inculpations, for example) than a strategy – available to be used for particular purposes in particular contexts, and varying in salience and in character according to when and how it is applied. Neither the Yugoslav nor the English writers discussed here are free to invent their own identities – they are constrained by the social and ideological resources they draw upon – but they are *makers* and *users* of difference as well as its victims. Thinking about the ways these writers used notions of masculinity and gender also helps move us beyond generalizations about the 'feminized other' and 'Western hegemony' based on monolithic and ahistorical concepts of both gender and power to an appreciation of the varied and changing ways in which systems of difference can interact, and at the same time to a more complex understanding of the character of East/West divisions within Europe.

Wendy Bracewell

Wendy Bracewell is Professor of Southeast European History at UCL's School of Slavonic and East European Studies. She has published extensively on travel writing by east European authors, and headed a major research project on the topic (with three volumes published under the series title 'East Looks West', CEU Press, 2008-09).

Notes

This research was supported by an AHRC Research Leave Award, and was carried out in the framework of the AHRC Research Project 'East Looks West'. I have benefited a great deal from comments by Bob Shoemaker, who knows a thing or two about gender, power and history.

1. This was neither a new mode of writing travel, nor one limited to writing on the Balkans: Holland and Huggan (2000: 27–37) discuss imperialist nostalgia and the 'English gentleman traveller' with particular reference to Eric Newby and Redmond O'Hanlon.
2. For further discussion of gender models in the Balkans in historical perspective, see Jovanovic and Naumovic (2004).

References

Antohi, S. 2002. 'Romania and the Balkans: From Geocultural Bovarism to Ethnic Ontology', *Tr@nsit-Virtuelles Forum* 21, http://www.iwm.at/t-21txt8.htm. Accessed June 2007.
Bakic-Hayden, M. 1995. 'Nesting Orientalisms: The Case of Former Yugoslavia', *Slavic Review* 54:4, 917–31.
Behdad, A. 1994. *Belated Travelers: Orientalism in the Age of Colonial Dissolution*. Durham, NC: Duke University Press.
Bjelic, D. and O. Savic, eds. 2002. *Balkan as Metaphor: Between Globalization and Fragmentation*. Cambridge, MA: MIT Press.
Blunt, A. 1994. *Travel, Gender and Imperialism: Mary Kingsley and West Africa*. London: Guilford Press.
Bohls, E.A. 1995. *Women Travel Writers and the Language of Aesthetics, 1716–1818*. Cambridge: Cambridge University Press.
Carr, H. 1985. 'Woman/Indian: "The American and his Others"', in *Europe and its Others: Proceedings of the Essex Conference on the Sociology of Literature*, ed. F. Barker et al. Colchester: University of Essex.
Carver, R. 1998. *The Accursed Mountains: Journeys in Albania*. London: John Murray.
Connell, R.W. 1995. *Masculinities*.Cambridge: Polity Press.
Crane, N. 1996. *Clear Waters Rising*. London: Viking.
Dimic, M. 1983. *Monah čeka svoju smrt: proza, zapisi, epifanije*. Priština: Jedinstvo.
Džadžic, P. 1987. *Homo Balcanicus, Homo Heroicus*. Belgrade: BIGZ.
Fanon, F. 1967. *Black Skin, White Masks*. New York: Grove.
Garton Ash, T. 1993. *The Magic Lantern: The Revolution of '89 Witnessed in Warsaw, Budapest, Berlin, and Prague*. New York: Vintage.
Goldsworthy, V. 1998. *Inventing Ruritania: The Imperialism of the Imagination*. New Haven: Yale University Press.
Goodwin, J. 1994. *On Foot to the Golden Horn: A Walk to Istanbul*. London: Vintage.

Hawks, T. 2000. *Playing the Moldovans at Tennis*. London: Ebury.
Holland, P. and G. Huggan. 2000. *Tourists with Typewriters: Critical Reflections on Contemporary Travel Writing*. Ann Arbor: University of Michigan Press.
Jovanovic, M. and S. Naumovic, eds. 2004. *Gender relations in South Eastern Europe: Historical Perspectives on Womanhood and Manhood in 19th and 20th Century*. Münster: Lit.
Kabbani, R. 1994. *Imperial Fictions: Europe's Myths of Orient*. London: Pandora.
Kaplan, R. 1993. *Balkan Ghosts: A Journey Through History*. New York: St. Martin's.
Kapor, M. 1979. *Skitam i pričam: putopisni dnevnik*. Belgrade: Prosveta.
Kiossev, A. 2002. 'The Dark Intimacy: Maps, Identities, Acts of Identification', in *Balkan as Metaphor: Between Globalization and Fragmentation*, ed. D. Bjelic and O. Savic. Cambridge, MA: MIT Press.
Konstantinov, A. 1966 [1895]. *Bai Ganio*. Sofiia: Bŭlgarski pisatel.
Kušan, I. 1986. *Prerušeni prosjak*. Zagreb: Znanje.
Lawrence, K. 1994. *Penelope Voyages: Women and Travel in the British Literary Tradition*. Ithaca: Cornell University Press.
Leigh Fermor, P. 1986. *Between the Woods and the Water*. London: John Murray.
Lewis, R. 1996. *Gendering Orientalism: Race, Femininity and Representation*. London: Routledge.
Lowe, L. 1991. *Critical Terrains: French and British Orientalisms*. Ithaca, NY: Cornell University Press.
MacLean, R. 1992. *Stalin's Nose: Across the Face of Europe*. London: HarperCollins.
McClintock, A., A. Mufti and E. Shohat, eds. 1997. *Dangerous Liaisons: Gender, Nation, and Postcolonial Perspectives*. Minneapolis: University of Minnesota Press.
Miller, H. 1958. *The Colossus of Maroussi*. New York: New Directions.
Mills, S. 1991. *Discourses of Difference: An Analysis of Women's Travel Writing and Colonialism*. London: Routledge.
Močnik, R. 2002. 'The Balkans as an Element in Ideological Mechanisms', in *Balkan as Metaphor: Between Globalization and Fragmentation*, ed. D. Bjelic and O. Savic. Cambridge: MIT Press.
Petkov, K. 1997. *Infidels, Turks, and Women: The South Slavs in the German Mind, ca. 1400–1600*. Frankfurt am Main: Peter Lang.
Roessel, D. 2002. *In Byron's Shadow: Modern Greece in the English and American Imagination*. Oxford: Oxford University Press.
Said, E. 1978. *Orientalism*. London: Routledge and Kegan Paul.
Schick, I. 1999. *The Erotic Margin: Sexuality and Spatiality in Alteristist Discourse*. London: Verso.
Spivak, G. 1988. 'Can the Subaltern Speak?', in *Marxism and the Interpretation of Culture*, ed. C. Nelson and L. Grossberg. Urbana: University of Illinois Press.
Stoler, A.L. 1991. 'Carnal Knowledge and Imperial Power: Gender, Race and Morality in Colonial Asia', in *Gender at the Crossroads of Knowledge,* ed. M. di Leonardo. Berkeley: University of California Press.
Storace, P. 1996. *Dinner with Persephone*. New York: Pantheon.
Thompson, M. 1992. *A Paper House: The Ending of Yugoslavia*. London: Hutchinson Radius.
Todorova, M. 1997. *Imagining the Balkans*. New York: Oxford University Press.
Whittell, G. 1992. *Lambada Country: A Ride Across Eastern Europe*. London: Chapmans.
Wolff, L. 1994. *Inventing Eastern Europe: The Map of Civilization on the Mind of the Enlightenment*. Stanford: Stanford University Press.

Chapter 7

AMONG CANNIBALS AND HEADHUNTERS
Jack London in Melanesia

Keith Newlin

In April 1907 Jack London set sail from San Francisco on a projected seven-year, round-the-world voyage. He spent five months in Hawaii, where he completed the outfitting of his boat, the *Snark,* while also touring the islands and writing a number of essays that financed his trip. After a difficult traverse to Nuku-Hiva, where he sought out locations memorialized by Melville's *Typee,* he set sail for Tahiti, prepared to see the tropical paradise promoted by the various travelogues he had read prior to his voyage. After wandering among the islands of Bora Bora, Samoa, Fiji, and Tanna, he landed in Guadalcanal in July 1908, which he used as a base to explore neighboring islands before illness caused him to end his voyage in December 1908.

Since he had arrived at Nuku-Hiva in early 1908, London and his wife, Charmian, had been systematically buying artworks from various islands and then shipping them home, and during the five months he spent in Melanesia in the summer of 1908 he eventually acquired some thirteen crates of objects that he sent to his ranch in Glen Ellen. Among the items listed on the manifest for the cases are 118 spears; forty-two paddles; 240 strings of shell money; dozens of carefully itemized shells, fishhooks, clothing, and shell jewelry; and bundles of shields, war clubs, axes, and other implements of war.[1] Curiously, there is a notable absence of the single-most important objective of that warfare in the Solomon Islands: there are no skulls. Nor are there any photographs of skulls in the massive archive of photographs that London assembled during his voyage. London had traveled to the Solomons with two related objectives: he wished to see first-hand the headhunting and cannibalism that served to mark the Solomons in the popular imagination, and then he wanted to turn that personal observation into essays and fiction for the magazines that underwrote his voyage.

For a number of scholars the *Snark* voyage has been regarded as responsible for a shift in London's thinking about race. In *Jack London's Racial Lives,* for example, Jeanne Reesman argues that the voyage "occasioned a dramatic change

Notes for this chapter begin on page 126.

in his racial thinking, as he learned much more about the diverse peoples of the world than his earlier racialist ideas had allowed," and that his "racially progressive stories sharply critique the social Darwinist idea of the 'inevitable' white man and Western misconceptions about tropical peoples" (2009: 109). This seems an appropriate way to characterize London's stories set in Polynesia, particularly those, such as "Koolau the Leper," "The Chinago," and "Chun Ah Chun," whose plots cast white colonizers as oppressors of peaceful citizens, but, I shall argue, that view does not hold for his attitude toward Melanesians. As Reesman admits, "He usually could not get past the darker skin and way of life in the Solomon Islands" (2009: 109), although stories like "Mauki," perhaps London's most violent and apparently racist story, she argues, "deserves close attention to its satire of racism," for it is "deliberate in its racial transgressions" (2009: 146). In that story, Mauki, an exploited and viciously treated Solomon Islander, takes his revenge against a cruel overseer by torturing him before taking his head. As other scholars have noted, London's attitude—his admiration for the lighter-skinned Polynesians and his disdain for darker-skinned Melanesians—is in keeping with racialist and ethnographic theories of human development, an attitude navigator Jules-Sébastien-César Dumont d'Urville influentially reflected in an 1832 article in which he divided the Pacific into the regions of Polynesia, Micronesia, and Melanesia (accompanied by a map representing the division still in use today) and where he characterized Polynesians as marked by civilization and beauty and the darker-skinned Melanesians "as degraded by the state of barbarism" (Thomas 1989: 30).[2]

Later on, I'll be suggesting an alternative perspective, one that questions how enlightened these stories—and London's attitude—actually are. But to get there, we first need to puzzle over two related conundrums: the first is that given the wide perception of the dangers presented by Melanesian headhunting, cannibalism, and seemingly indiscriminate violence, why on earth would London put his life as well as the lives of his crew at risk? As he wrote to George Sterling while at Guadalcanal on October 31, 1908: "This is about the rawest edge of the world. Head-hunting, cannibalism and murder are rampant. Among the worst islands of the group and night and day we are never unarmed, and night watches are necessary" (1988: 2:770). In the preface to *The Cruise of the Snark,* London explained that his motive for undertaking the *Snark* voyage was primarily the pursuit of adventure, one in which he could measure his ability to adapt to a challenging environment: "The achievement of a difficult feat is a successful adjustment to a sternly exacting environment" (1911b: 5). And as David Farrier has argued, "London's adventure was, primarily, an act of will, an assertion of his capacity to go and do as he pleased in the tradition of the American frontiersman" (2007: 178)—and the frontier had shifted to the South Pacific. Perhaps London's clearest expression of his attitude appears in the mouth of the character Joan Lackland in the novel *Adventure*: "But to be among them [the cannibals of Guadalcanal, the setting of the novel], controlling them, directing them, two hundred of them, and to escape being eaten by them—that, at least, if it isn't romantic, is certainly the quintessence of adventure" (1911a: 58). London's voyage into looming danger, in a region widely regarded as "the rawest edge of the world," became a means to verify himself as a masculine adventure-hero in a savage landscape that would test his courage and ability to navigate danger. As Paul Lyons has remarked, "the discourse of cannibalism, with its thrilling gaze at (and into) the 'nature' of others, thus functions as a sign for tourism itself—for packaged escape into and out of a zone of difference that functions to reify the gazer's worldview" (1995: 42).

The second conundrum concerns the dramatic highlight of the Melanesian portion of his voyage, the problem presented by the grounding of the blackbirding ship *Minota,* an event in which London faced imminent death at the hands of angry islanders. For London was no mere passive participant: he eagerly sought out the opportunity to experience blackbirding—the coercive recruitment of indigenous peoples to work as laborers—and more: when he learned that the blackbirding ship *Eugenie* was on a reef, he offered the use of the *Snark* as a substitute.[3] The conundrum is that London had made his reputation agitating for the rights of labor, and indeed, as Charmian London reports in *The Log of the Snark,* he had delivered his lecture "Revolution" on May 12, 1908, in Samoa (1915: 271), just three months before. In that lecture, after presenting a number of examples of worker exploitation and an argument that revolution against exploitive capitalists would ultimately prevail, he concluded, "The revolution is here, now. Stop it who can" (2008: 156). How could he reconcile his argument to stop the exploitation of laborers by the capitalists with his own participation in labor exploitation?

London and Travel Writing

To answer these questions, we first need to detour into the travel writing London had absorbed prior to and during his voyage. This writing, especially descriptions of cannibalism and headhunting that served to mark the savage Other, shaped London's expectations for what he would encounter, and it would later shape his own travel account, the essays he wrote during the voyage and later collected into *The Cruise of the Snark,* as well as the fiction that emerged from his experience and especially the photographs he selected to illustrate his writing.

About the same time as London's voyage, journalist Beatrice Grimshaw traveled to the South Seas, writing publicity for some of the islands and reporting on events for the newspapers. She wrote a number of books and articles about her adventures that did much to shape public perceptions, as in this excerpt from *In the Strange South Seas* (1907), published in *National Geographic* magazine:

> Westward of the Fijis lie the dark, wicked, cannibal groups of the Solomons, Banks, and New Hebrides, where life is more like a nightmare than a dream; murder stalks openly in broad daylight, people are nearer to monkeys than human beings in aspect, and music and dancing are little practiced and in the rudest possible state. (1908: 2)[4]

Here Grimshaw draws a pointed contrast to the peoples of Polynesia, whom travelers had long admired for their beautiful physical form and accomplished music and dance and whose seductive hula dancing and nudity had long beguiled travelers.

When he dipped into his copy of Charles M. Woodford's *A Naturalist Among the Headhunters* (1890), London read this frightening remark:

> Head-hunting to a greater or less extent is carried on by most of the inhabitants of the Solomon group, but it is from New Georgia and the adjacent islands that it is most extensively indulged in. During a former visit of a fortnight to the Rubiana lagoon no less than thirty-one heads were brought home to different villages round the lagoon and islands near. With these natives it appears to be a perfect passion. (1890: 153)

Woodford, whom the Londons would meet on Tulagi (then the capital of the Solomon Islands Protectorate) and with whom they had many conversations, believed that the islanders were "savages," in the sense of primitive, violent, and uncontrollable beasts, and he warned his readers,

> From my somewhat wide and varied experience of them, I am of opinion that the first thought that animates a native upon the sight of a stranger is, "Will he kill me?" Having answered this to his own satisfaction, his next thought is, "Can I kill him?" the latter question being considerably influenced by the fear of future retribution to be apprehended from the friends of the stranger, in case he is a native.

But, he concluded, "in the case of white men this fear of retribution hardly enters as a factor" because of "an absence of the fear of future consequences, while the hope of plunder is doubtless in many cases sufficient incentive" (1890: 32).

And from his copy of Herbert Caley-Webster's *Through New Guinea and the Cannibal Countries* (1898), he read additional confirmation of the portrayal of Melanesians as treacherous, animal-like savages:

> These natives are not only head-hunters and cannibals, but make no secret of it whatever. They are the most treacherous of all the people of the South Seas, and when apparently on the most friendly terms are only awaiting a favourable opportunity to catch the stranger unawares, and to add one more head to their already huge collection. I may say that during the whole of my visit I hardly ever had my revolver out of my hand. (1898: 108)

Because his reading had prepared him to expect to encounter islanders with little regard for human life—indeed, who apparently chopped off heads for the sheer animal thrill of it—London came to the islands well-armed. Among the items shipped home when he abandoned the voyage are two .22 Marlin rifles, three .32 Winchester rifles, one Mauser rifle, a Remington autoloading rifle, two shotguns, and four revolvers—plus some 462 boxes of shells and an unopened case of .22 rifle cartridges.[5] This, by the way, is what is left over from the many rounds the Londons fired to intimidate the natives, as when, upon arrival at Ugi (Uki Ni Massi), a small island north of San Cristobal on July 4, Charmian reports that "we haled forth every dispensable bottle, match-box, piece of cardboard, cocoanut shell, and went at a demonstration of marksmanship that ought to make us taboo from any 'monkeying' in these parts. Mausers, automatic rifles, Colt pistols, Smith & Wesson revolvers, and Mr. Hammond's Sniders, all proved whether or not they were rusty" (1915: 362).

The Londons' fear of the Melanesian Islanders is clearly apparent in all three accounts of the voyage.[6] In *Through the South Seas with Jack London*, Martin Johnson describes their first landfall at Port Mary on Santa Ana island on June 28, when, upon anchoring,

> we were surrounded in an incredibly short time by a hundred canoe-loads of savages—people who in looks and actions fully justified my expectations of what South Sea Islanders should be like. They started aboard, but with guns we kept them back; and they circled round the *Snark*, waving spears and clubs and shouting at the top of their voices. And their looks certainly made necessary the precautions we took, for they were a most savage-looking lot. (1913: 287)

As Johnson makes clear, his fear stems primarily from the islanders' "savage-looking" appearance, whose "looks and actions fully justified" the expectations created by his reading. In this instance, however, the savage look was deceiving, for the islanders who surrounded the *Snark* were a delegation sent by a local trader to welcome the London party (C. London 1915: 347).

As it turns out, despite registering their fear of the Melanesians throughout their accounts, none of the London party encountered any cannibals or headhunters; what they did encounter was their own fear: fear of the savage Other, fear of the alien, a terror that had been whipped up by their reading and the tales told to them by the traders and planters they met with during the voyage.[7] And the reason for their failure to find what they had traveled specifically to see—evidence of cannibalism, eye-witness observations of headhunting in action—is that neither was present in Melanesia at the time of their voyage.

Cannibalism and Headhunting

Cannibalism in Melanesia had never been as widespread as depicted, if it even existed at all. As William Arens argued in his controversial book *The Man-Eating Myth* (1980), there is little empirical evidence of actual cannibalism; rather, cannibalism is largely a projection by Europeans who sought to justify their exploitation of indigenous people by depicting them as the savage Other, as animals beyond the pale of civilization. More recently Gananath Obeyesekere, in *Cannibal Talk* (2005), has extended Arens's thesis in a deconstructive reading of South Seas travelers' tales to argue that cannibalism is mostly "talk" by both colonizers and the colonized, intended both to intimidate and to defend against exploitation. And in an influential study of Roviana predatory headhunting, Shankar Aswani argues that part of the headhunting ritual in the Solomon Islands entailed objectifying victims by stripping them of personhood through representing them as animals, as fish or pigs (2000: 56) and that their talk "metamorphosed humans into 'wild' fish and animal prey. . . . Human prey was also depicted as slaughtered pigs" (2000: 62). Moreover, to prevent victims' angry spirits from taking revenge, warriors would burn pieces of the victim's body as an offering at a shrine (2000: 58). It's easy to see that in their early encounters Europeans likely misinterpreted the metaphorical as literal, misperceiving talk about hunting, cooking, and eating pig to mean the hunting and literal consumption of humans.[8]

One fact emerges: London never witnessed a "cannibal feast" as he hoped, despite much eager searching. Charmian records one instance of hearing talk from an elderly islander about past glory days of headhunting raids and cannibalism. "He told me with crackling glee and horrible grimaces, of the numerous white men he had killed in his day. . . . But you cannot get any of them to admit they have 'kai-kai'd' human flesh. They know our abhorrence of this practice, and look sheepish and silly when questioned directly" (1915: 363). This pattern—verbal reports, often told with exaggeration and careful attention to response—marked the Londons' experience of cannibal talk, as it had for the travelers before them. So important was attaining empirical evidence of cannibalism that Martin Johnson returned to the region in 1917 with the specific aim of filming a cannibal feast to enthrall audiences in his vaudeville stage shows and films. With his financial backers expecting visual evidence, Martin searched assiduously but was unable to find evidence of cannibalism.[9]

Nor did London witness any headhunting raids, although his fear of headtaking dominated his essays about Melanesia, as it also did in Charmian's and Martin's parallel accounts. And the reason is that headhunting had largely been suppressed by 1900.

For generations before encounters with Europeans, inhabitants of the islands centered around New Georgia had practiced taking heads in various intergroup conflicts. Political groupings were small and tended to be organized around clans and some clan alliances, and leaders (*banara*) emerged who could "bring renown to himself and his supporters by conspicuous demonstration of wealth and, above all, his generosity" (Bennett 1987: 14). Interclan warfare tended to arise from raids to steal such resources as material wealth and captives, especially women, which the banara would then redistribute to maintain his leadership. For the islanders, the head was the seat of *mana*, the life force. Taking the head of an enemy, particularly the head of a leader or prominent warrior, deprived that enemy of power and respect (Aswani 2000: 61), for without the head, the disembodied *mana* "created harmful forces within the enemies' communities ... and prevented the transformation of their dead into viable ancestors" (Dureau 2000: 82).[10] As Deborah Waite explains, "leaders gained prestige and power through organising and participating in successful head-hunting raids. Heads obtained on raids attested to the power of leaders, most particularly, their effective leadership ability facilitated, in part, by success in communicating with spirits that made power operative" (2000: 124). The headhunting parties would attempt to kill all members of the target village to decrease the likelihood of revenge killings, which need not be against the perpetrator—it was sufficient to take revenge against any member of the group.[11]

These trophy heads were then subjected to ritualized degradation to further alienate the now-departed *mana*. As Aswani describes it,

> Once cut, the heads of victims were smoked and rinsed, and then carried back to the village "like a bundle of coconuts." Following the return of a successful expedition with the blow of a conch shell (*buki hogoto*) and the warriors' welcome with dance and songs by an expectant village (*peka*

aqa), the victims' skulls were placed in a tabernacle (*patu kevuana*) in the centre of a dancing ground (*pavasa*). A ritual specialist would spit on the skulls a mix of betel nut and ginger leaf, and recite a particular secret charm to "cool down" (*ya-ibu-a*) the revengeful spirits of the victims. In addition, an offering would be made at an ancestral shrine to celebrate the expedition's safe return. (2000: 63)

To demonstrate their success in battle, victorious raiders would display the heads of prominent victims in canoe houses, and the heads were often decorated: "Shell rings could be bound to the skull or it could be dried, modelled with *Parinarium* nut paste, painted black and inlaid with shell, with a resultant appearance" that often appeared life-like (Waite 2000: 124). But those of ordinary warriors were simply buried.

Travelers to the region also encountered skull shrines (*zelepade*) in which the skulls of ancestors were kept and venerated. While one of the purposes of predatory headhunting was to deprive enemies of *mana* and so contribute to the disarray of their enemies, skulls of ancestors were ornamented with shell rings, and entrance was forbidden to all but mortuary priests who attended to the shrine and presided over ritual offerings (Waite 2000: 125–126). One can readily imagine the horror such shrines and canoe houses presented to travelers who had little understanding of the cultural background of such displays so foreign to European experience, as in this report by artist Norman H. Hardy, who traveled to the Roviana lagoon and apparently met Ingova, the most notorious of the headhunting leaders. Hardy described Ingova's *paele,* or canoe house:

> Years ago Ingova's [*paele*] was hung with skulls, hundreds of them were strung in the cross-beams, with staring, vacant eyeholes, which looked out of nothing and yet seemed to see everything. Their drooping lower jaws, showing sets of white teeth which glistened in the rays of the moon, made Ingova's heart throb with pride as he stood and tried to count them. White naked skulls of brave men all hung in rows. (Elkington 1907: 98)[12]

In 1892 the British government established a protectorate in the Solomons to freeze out other colonial powers but decreed it must be self-financing. Charles Woodford was made acting resident commissioner in 1896 and quickly saw that the only way to finance the new protectorate was to establish plantations, which would mean alienating land from the indigenous population. To attract the necessary capital, pacification would be necessary. Because he had only eight Fijian policemen to enforce the new colonial order, Woodford enlisted friendly clans to fight unfriendly groups, and he also enlisted resident traders as a sort of militia (Bennett 1987: 106). But mostly he relied on British naval warships to shell villages in what Judith Bennett calls "massive overkill" (1987: 109). As Aoife O'Brien explains,

> colonial police frequently targeted objects held in great esteem by Indigenous people. Canoes and canoe houses, valuable heirloom objects, and other cosmological and materially valuable items, such as human heads, were alternatively destroyed or collected by colonial officials and their police forces. (2017: n.p.)

The effect was to eradicate native culture, which is one reason the London party was briskly buying up native artifacts, for the colonial powers widely believed that the islanders were doomed to extinction (see Bennett 1987: 125). As plantations became established and copra production increased, along with trade, the effect was to decrease the power of *banara* and their role in trade and alliances (Bennett 1987: 112–113). By 1900 Woodford's shock-and-awe tactics had "stopped headhunting from Roviana, Simbo, and Mbilua and enforced peace among adjacent peoples" (Bennett 1987: 107).

So when London arrived at Penduffryn Plantation on Guadalcanal in the summer of 1908, he saw no headhunting, despite looking specifically for it, because it had ceased. He took no pictures of skulls in part because many sites had been destroyed and because, for the few that remained, of the limits of photographic technology that made indoor shots difficult—and because taking the photos would have required permission from the islanders, who would not permit the entrance of foreigners whose presence would desecrate shrines.[13] As one scholar remarks, "In the absence of direct evidence, photographs of Roviana warriors holding spears or axes and wicker shields, of the large canoes (*tomoko*) used in headhunting raids (but also for trading expeditions), often operated as visual stand-ins for headhunting" (Wright 2013: 20). This is a strategy London himself adopted in his photographs depicting native peoples with captions such as "Man Eaters" and "Cannibal Bushmen at Forte, Northwest Malaita" (see illustrations 7.1 and 7.2).

In illustration 7.1 a group of men sit in boat, staring directly at the camera, a gun hoisted over a shoulder, a bundle of spears held by a hand. In illustration 7.2 a group of men sit on a log, a rifle prominently held at center, their arrangement suggesting a posed shot. Are these cannibals or people gathered to be photographed? As Susan Sontag has suggested, "There can be no evidence, photographic or otherwise, of an event until the event itself has been named and characterized. And it is never photographic evidence which can construct—more properly, identify—events; the contribution of photography always follows the naming of the event" (1973: 19). With the captions London has named the

Among Cannibals and Headhunters

Illustration 7.1: "Man-Eaters." From *The Cruise of the Snark* (1911), 260.

Illustration 7.2: "Cannibal Bushmen at Forte, Northwest Malaita." From "Cruising in the Solomons," *Pacific Monthly* 23 (June 1910), 593.

"event"—the cannibal—and the details of composition affirm what London has led his readers to expect in his text: the naked torsos emphasizing musculature, the subjects staring directly into the camera with weapons displayed, their ears and noses pierced, the outward signs of the savage.

What the Londons did witness was the islanders' violent response to colonial labor practices. Labor recruiting before the turn of century focused on recruiting laborers for Fiji and Queensland sugar fields, leading to population declines and employment of coastal dwellers as middlemen, who facilitated the acquisition of bushmen, who had more difficulty in wresting a living from the mountainous land than did coastal dwellers, especially from Malaita and Guadalcanal (Cooper 1979: 36). Scholars estimate that between 1863 and 1914 about one hundred thousand Melanesians were recruited as indentured laborers in a system that was ripe for exploitation and abuse (as reflected especially in the story "Mauki"), despite several regulations being enacted to police the trade. Abusing laborers was common. As Bennett summarizes, beatings were routine:

> Planters regularly gave laborers a blow over the ear or a kick to the backside to make them do what they were told. Violence was so much a part of the expected disciplinary methods of the planters that a white man had to prove himself competent in it to be a success. It was common practice to ask overseers and managers if they could fight. (1987: 169–170)

In 1901 Queensland voted to enact a new law to provide for the "Regulation, Restriction and Prohibition of the Introduction of Labourers from the Pacific Islands"—basically an act to expel nonwhite residents of Queensland as part of a movement to maintain white hegemony. One of the provisions stipulated that, by 1906, all Melanesian laborers were to be deported back to their island of origin (Docker 1970: 260). In 1906 there were 6,389 Melanesians subject to the new deportation order; of these, about 4,800 were from the Solomons, with 2,500 from Malaita, the most populous island (Cooper 1979: 49). Repatriated laborers, most of whom were bushmen who had been attracted to plantation service to improve their lives, both unified and divided Malaitan society: their labor abroad had effectively reduced linguistic and political divisions while it also augmented their desire for European goods, yet they also posed a grave threat to the old order: they had become independent, relatively prosperous, and powerful because of smuggled firearms, so they were no longer dependent on *banara* for support as they had been previously (Cooper 1979: 51). Malaitans also had the reputation of being the most savage of the islanders, "fractious, volatile, and violent ... self-assertive, truculent, and contemptuous of others" (Cooper 1979: 48). In 1906, when Malaitans began to return, there was no government station on Malaita, pacification efforts had been only partially effective, and the only resident Europeans were a few missionaries. As one scholar put it, "the island which was destined to bear the brunt of the repatriation was 'precisely the one least equipped to bear it without violence and disruption'" (Cooper 1979: 49).

So when London sailed into the harbor at Su'u on the west coast of Malaita on August 8, he entered a region rife with internecine conflict before the British exerted control over the region with the establishment of a government station in 1909 (Cooper 1979: 57), after London had abandoned his cruise. While at Penduffryn Plantation, Captain Jansen of the *Minota* had invited the Londons to accompany him as he ferried a cargo of returnees to Malaita and then picked up a new batch of recruits for plantation labor on Guadalcanal. (While the trade in laborers to Queensland had ended, a brisk trade was still underway for plantations within Melanesia.) Just six months before, in December 1907, the *Minota* had arrived in Alite, Malaita, to recruit laborers for work on Guadalcanal plantations, under the command of an inexperienced captain named C. C. MacKenzie. As Woodford explained in a letter to Everard im Thurn, the governor of Fiji, the "mere fact of sending an inexperienced white man ... to Malaita was sufficient to issue disaster" (qtd. in Cooper 1979: 55). Bushmen attacked and killed MacKenzie, and "the local saltwater men, hearing the disturbance, came aboard the ketch, rescued the Solomon Islands crew, and sailed the vessel back to Florida Island, before reporting the murder to government headquarters at Tulagi" (Cooper 1979: 55). In response, Woodford sailed on the HMS *Cambrian* and, with a company of soldiers, "landed at Bina, going inland to destroy the villages of those implicated" (Bennett 1987: 109). As it turns out, London had journeyed up the coast on the *Minota* and arrived at Langa Langa lagoon just after the *Cambrian* had departed that morning. Later that day he learned from a missionary that while "the villages had been burned and the pigs killed ... the murderers had not been captured" (London 1911b: 271–272). Given the punitive raid, islanders understandably refused to be recruited. "Three fruitless days were spent at Su'u," London recorded. "The *Minota* got no recruits from the bush, and the bushmen got no heads from the *Minota*" (1911b: 270–271). Charmian writes on August 13 that "As there was no chance of gathering any recruits from the troubled bush region, we set out for Malu, on the north side of the island, to land the last of the homing blacks and drum up a new supply" (1915: 401).

Two days later, at Malu Bay, the *Minota,* with its decks surrounded by a double row of barbed-wire fencing to keep marauding islanders at bay, at last succeeded in convincing some Malaitans to sign up for labor in Guadalcanal (C. London 1915: 404). On August 19, with a cargo of forty recruits aboard, the *Minota* set sail for Guvutu, on Florida Island, to meet a steamer and get their mail. But the *Minota* grounded on a reef. In their respective accounts of what transpired next, as part of his self-promotion as an adventurer-hero, London emphasized the resultant panic and fear—fear, he implies in *Cruise,* of attacks by headhunting Malaitans eager to secure trophies:

> Bedlam reigned. All the recruits below, bushmen and afraid of the sea, dashed panic-stricken on deck and got in everybody's way. At the same time the boat's crew made a rush for the rifles. They knew what going ashore on Malaita meant—one hand for the ship and the other hand to fight off the natives. ...
>
> When the *Minota* first struck, there was not a canoe in sight; but like vultures circling down out of the blue, canoes began to arrive from every quarter. The boat's crew, with rifles at the ready, kept them lined up a hundred feet away with a promise of death if they ventured nearer. And there they clung, a hundred feet away, black and ominous, crowded with men, holding their canoes with their paddles on the perilous edge of the breaking surf. In the meantime the bushmen were flocking down from the hills, armed with spears, Sniders, arrows and clubs, until the beach was massed with them. (1911b: 287–288)

But in her diary Charmian makes clear that what the Malaitans were really after was salvage: the *Minota* was laden with trade tobacco in addition to recruits, so they weren't exactly in fear of their lives but were fearful of being looted. "They do not necessarily kill," she explained, "unless a good opportunity offers, but are mad for the loot" (JL 207, Jack London Papers). The day before, the Londons had sought to intimidate the islanders by a shooting demonstration. "At Jansen's sociable suggestion"—Charmian writes in *The Log of the Snark*—"as if for the special entertainment of the others, Jack emptied a few magazines from his Colt's Automatic, and the bushmen stared and emitted guttural sounds of astonishment and awe at the stream of lead the pickaninny fella gun belong white man could pour out." Not to be outdone, Charmian also demonstrated her marksmanship: "My modest Smith & Wesson, being in the hands of a mere Mary, impressed them to foot-shifting embarrassment. The fact that we can hit objects at a distance also acts as a check to undue mischievousness on their part" (1915: 411–412). So the Londons knew exactly how dangerous the recruiting trip was, but they apparently took comfort in the *Snark*'s arsenal as well as the perception that, although murderous, the natives were innately cowardly and easily cowed by the shock-and-awe tactics of the recruiters and traders.

London as Tourist

With this as a backdrop, let us return to the two conundrums I presented earlier. Why place himself and his crew in danger? Although London did recognize the exploitation of islanders and does take the viewpoint of Melanesians in many of his stories by presenting the effects of colonial labor practices, he also saw them as lesser human beings, people easily dominated by the superior intelligence and armament of the colonizers. This perception appears most clearly in his descriptions of body piercings, which he never apparently makes an effort to understand; rather, his comments are condescending, and he holds the cultural practice up for ridicule—a mark of his position as an "inevitable white man." "To look at, they were certainly true head-hunting cannibals," London begins a typical description, thus revealing how his reading has prepared him to see Melanesians.

> Their perforated nostrils were thrust through with bone and wooden bodkins the size of lead-pencils. Numbers of them had punctured the extreme meaty point of the nose, from which protruded, straight out, spikes of turtle-shell or of beads strung on stiff wire. A few had further punctured their noses with rows of holes following the curves of the nostrils from lip to point. Each ear of every man had from two to a dozen holes in it—holes large enough to carry wooden plugs three inches in diameter down to tiny holes in which were carried clay-pipes and similar trifles. In fact, so many holes did they possess that they lacked ornaments to fill them; and when, the following day, as we neared Malaita, we tried out our rifles to see that they were in working order, there was a general scramble for the empty cartridges, which were thrust forthwith into the many aching voids in our passengers' ears. (1911b: 264–265)

In this description the image conforms to the type presented by earlier literature, for the mark of the savage Other, "the true head-hunting cannibal," are the multiple piercings so alien to European standards of beauty—who else but a cannibal would perforate nostrils and ears and fill the "aching voids" with objects more suited for pockets? Only naked savages, London suggests, who resort to making pockets of the skin: their cleverness also suggesting the wily, deceptive nature of the man-eating savage.

Charmian displays her racial arrogance even more clearly in the *Log of the Snark,* where, at Port Resolution on the island of Tanna in Vanuatu, in an entry for June 12, she records her impressions:

> And the natives: As I write, near by but not too near (they may be clean but they don't look it), squat a half-dozen of the strangest human beings I ever beheld outside a feeble-minded institution. We had heard they were the lowest of the Melanesians, but they excel all expectations. Bodies are thin and unbeautiful, with bulges in the wrong places; legs show thin and crooked, and their generally evil, low-browed malformed Black-Papuan faces are curiously repulsive. One old fellow, a trifle less unpleasant than the rest, has an expression that is intended to be benevolent, on a nut-wrinkly face with unsecret, sky-turned nostrils, the eyes most remarkable with the vacillating intentness of a monkey, while he endeavours to compose his attention on the typewriter, at which I have been working on deck. He is quite the nearest to a chimpanzee that I've ever seen. The gaze focuses, wavers, comes back, and his lips narrow and widen with an undeveloped attempt at a human smile. The only way to fix an image like this, is to sit right down and write. (1915: 325)

To Charmian, Melanesians appear be only a step above the monkey on the chain of evolution, with limited intelligence, as suggested by bodies that reveal the effects of malnutrition and faces that present a "vast contrast to the chiselled heads of Fiji and Samoa," which more nearly approximate European standards of beauty (1915: 325). In her account, she frequently refers to Melanesians in animal terms, as "baboons," a "band of monkeys" with "monkeyish" smiles," or "animal-hairy humans," or more simply as "creatures" with "little minds." Part of the reason for her disdain is that Charmian, like many travelers before her, was enraptured with Polynesians, who had long represented an ideal of beauty and sexual licentiousness, and she was especially drawn to Polynesian art, as is reflected in her descriptions of trading for curios during that portion of their voyage. But in the more violent islands of Melanesia the Londons encountered a people still in the process of adapting to—and resisting—colonial exploitation, which was reflected in their general suspiciousness of strangers and also, Charmian believed, affected both their intelligence and their art:

> An old paralysed black heathen sat on the beach where we landed, and looked at me and my camera with sullen, unsympathetic gaze sans fear, sans interest, sans understanding, sans everything. It would seem that the only idea these people ever possessed was to kill. With that ambition quenched by the joint French and Australian colonies, they resolve into mere nonentities. Evidently all their craft went to the one passion; and their general lack of clever house-building or mat-weaving, or ornament-devising, would bear this out. (1915: 327)

The captions to the photographs that illustrate his travel pieces make London's objectification and condescension abundantly clear.[14] The caption to illustration 7.3, showing "Stone Fishing at Bora-Bora," reads, "A Native South Sea Island Missionary, in What He Considers Correct Clerical Garb." What London didn't realize was that natives adopted European dress when in the presence of Europeans in an effort to demonstrate equal status (see Bennett 1987: 97–98). The caption to illustration 7.4 echoes this condescension. Ironically, in the photo album Charmian assembled after the voyage, this photograph appears with her caption, "A Prince of Polynesia, Nuka Hiva, 1907" (figure 5), indicating a repurposing of the image from illustrating the beauty of Polynesians to signaling ridicule of Melanesians. And London's caption to figure 6, "The Two Handsomest Men in All the Solomons," where the direct, confrontational gaze of the subjects, their ornamental ear piercings, and their naked torsos confirm readers' expectations of images of savages, continues his pattern of holding black-skinned islanders up to ridicule.

London as Headhunter

London's voyage into the heart of Melanesia, though dangerous, was one he believed he could control: the islanders were perceived to be more animal than human, and humans had a long history of dominating animals and more "primitive" peoples. His regular demonstrations of superior firepower served to buttress his position as a superior human, and this attitude also perhaps explains his participation in blackbirding, where he basically functioned as an enabler. If he saw Melanesians as human beings with a dignity equal to that of Europeans, how could he participate in their exploitation?

In his travels through Melanesia Jack London operated as a metaphorical headhunter, collecting artworks and taking photographs, trophies to signify his presence in an exotic land and his skill at negotiation, just as Melanesian headhunters collected heads. Headhunters acquired heads in raids to demonstrate their

Illustration 7.3: "A Native South Sea Island Missionary, in What He Considers Correct Clerical Garb." From "Stone Fishing at Bora-Bora," *Pacific Monthly* 23 (April 1910), 342.

Illustration 7.4: "Laundry Bills Are Not Among His Vexations. His Garb, However, Is a Concession to Civilization.—Lord Howe Atoll. From *The Cruise of the Snark*, (1911), 332.

manliness and power as a warrior; their ability to acquire material goods to redistribute among their people proved their leadership and maintained order. London's acquisition of objects from a region perceived as violently dangerous is more than the tourist's penchant to document his tourism; the acquisition also suggests his power as a negotiator, his ability to navigate danger, and his manliness, and by documenting his observations with photographs as well as through display of art works, he similarly demonstrated his leadership.[15]

London's writing from Melanesia actually continues the tradition of depicting the "savage" Other and is not as innovative and humanistic as critics have argued. We can see this most clearly in his participation in labor recruiting and in his description of Solomon Islanders as well as in the photographs and captions that illustrated his writings. When London's personal experience of Melanesia did not give him the material readers had come to expect of South Seas fiction, he did what any fiction writer does: he invented episodes of cannibalism and headhunting for his fiction, and for his nonfiction he continued the pattern of depicting islanders as savages. In short, London's attitude toward the black islanders did not undergo a shift to perceiving their shared humanity; rather, it remained little different from what he had been reading.

Illustration 7.5: "A Prince of Polynesia," JLP 502, Alb. 64, Jack London Papers.

Illustration 7.6: "The Two Handsomest Men in All the Solomons," From "Cruising in the Solomons," *Pacific Monthly* 24 (July 1910), 38.

Illustration 7.7: "One Oreinte (Inca) Andes Indian reduced dried human." Courtesy of California State Parks, 2018.

Keith Newlin

There is a final codicil: appearing dead last on a list of curios Charmian later offered for sale is a curious item: "One Oreinte (Inca) Andes Indian reduced dried human head with long black hair. Unusually fine example, perfect. Valued at $350.00" (JL 317, Jack London Papers) (see illustration 7.7). Although the Londons did not find any heads in Melanesia, they did pick up one from Ecuador during their return voyage. Jack London could now claim to be a successful headhunter.

Acknowledgments

I am grateful for a UNCW Charles A. Cahill Award, which funded research in the Jack London Papers at the Huntington Library. I thank Sue Hodson, former literary manuscript curator at the Huntington, for her assistance with the London papers, as well as Clint Pumphrey, manuscript curator at the Merrill-Crazier Library at Utah State University, for making available copies of London materials held there, and Carol Dodge, of the California State Parks, for making available an image of the shrunken head London acquired on his way home from the *Snark* voyage. Thanks also to Anita Duneer and Heather Waldroup for their astute comments on an earlier draft of this essay.

Keith Newlin is Professor of English at University of North Carolina Wilmington, where he teaches courses in American literary realism and naturalism, modernism, and drama. The editor of *Studies in American Naturalism*, he is the author of *Hamlin Garland, A Life* (2008) and editor of *The Oxford Handbook of American Literary Realism* (2019), among other books. In addition to teaching as a Fulbright Senior Scholar in Germany, he has lectured widely in China on such topics as Jack London, Hamlin Garland, realism and naturalism, and literary authorship, as well as in Wales and South Korea.

Notes

1. The manifest is headed "Cases shipped by Jack London from Solomon Islands to San Francisco" and is accompanied by a letter, dated October 25, 1908, entitled "Manifest to Accompany Shipment of Curios," in which London gives examples of the exchange rate in trade tobacco (JL 317, Jack London Papers).
2. The translation is by Thomas, who in his article traces the history of anthropological distinctions between the two regions. See also Earle Labor, "Jack London's Symbolic Wilderness," for discussion of London's perception of Melanesia as an "Inferno" where "the Darwinian law operates in its most insidious forms" (1962: 152). For discussion of London's racism as reflected in his novel *Adventure* (1911a), see Philip Castile, "The Last Phase of the South Sea Slave Trade: Jack London's *Adventure*" (2012).

3. Charmian describes the offer in *The Log of the Snark*: while the *Snark* set out for Malaita with a cargo of returnees (along the way they encountered the *Eugenie* and learned that the story of its grounding was a joke by George Darbishire, one of the partners at Penduffryn Plantation on Guadalcanal), adverse winds sent them back to the plantation (1915: 382–383).
4. London's clipping file contains a copy of this article, although there is no indication of when he acquired it (see JL 1616, Jack London Papers). Subsequent references to books by Charles Woodford and Herbert Caley-Webster are also in the Jack London Papers, with a number of passages marked.
5. This is from a seven-page manifest ("*Snark* Invoices") listing the contents of twenty crates; the first page appears to be missing. The nature of the contents suggests London was shipping his personal belongings home after the sale of the *Snark* (London n.d.).
6. London's *The Cruise of the Snark* (1911b) is a compilation of the articles he wrote during the voyage to finance the trip; they are often humorous in tone and sometimes presented without regard to chronology. In addition, there are curious gaps: for instance, he writes nothing about his time in Fiji, during which he witnessed the eruption of a volcano that destroyed many of the villages, which would certainly have made good copy for the magazines. Charmian London's *The Log of the Snark* (1915) is a more traditional travel narrative, full of detailed descriptions and with many of the entries dated, reflecting its source in her diary. And Martin Johnson's *Through the South Seas with Jack London* (1913) is based on his diary and newspaper items he sent to Kansas newspapers; it was also ghost written by Ralph Harrison. For discussion of Charmian's book and its significance, see Tucker (2017).
7. For an illuminating historical account of how travelers expressed their fear of alterity in terms of racialized images, see Gustav Jahoda (1999), *Images of Savages*, especially part II: Animality and Beastly Man-Eating.
8. The literature devoted to the issue of the reality of cannibalism is extensive. As Jahoda points out, cannibalism "remains probably the most powerful symbol of savagery" (1999: 97) and serves as the chief image in the popular mind to represent the Other. Anthropologists continue to debate the existence of endo- , exo- , and ritual cannibalism— that is, cannibalism within one's community or outside of it, or for ceremonial purposes.
9. In *I Married Adventure,* Osa Johnson recounts that she and Martin stumbled across evidence of cannibalism on the island of Malekula, where their entrance into a clearing scared off islanders seated around a campfire. In the embers of the campfire they discovered "a charred human head, with rolled leaves plugging the eye-sockets." The films Martin took, she concluded, "proved conclusively that cannibalism there is still practiced" (1940: 160–161). But in a November 9, 1919, letter to Charmian, soon after the event, Martin describes the scene as a demonstration of head preservation: "They actually dried and smoked human heads before our own eyes, and I found the remains of a human head, cooked, and still hot on the fire, although the natives fled" (HM 473972, Jack London Papers).The Johnsons' biographers explain that when the Johnsons prepared the film for public presentation, they decided to characterize the curing ceremony as a "cannibal feast"—a characterization later repeated in their writings and picked up by commentators ever since (Imperato and Imperato 1992: 82–83).
10. In his revisionary interpretation of Roviana predatory headhunting, Aswani explains that the purpose of taking heads is not to absorb a victim's *mana*, or life force, as early ethnologists had claimed, but to create disarray among one's enemies: "By severing and taking heads and slaves, Roviana warriors abrogated the enemy's future benefits of securing the power of the enemies' own ancestors—the souls of many victims destined to become earthly prisoners seeing revenge against the living" (2000: 55).
11. Thus, when Europeans committed acts of violence against islanders, all Europeans could then become targets of revenge, which had the effect of increasing the violence.

12. As Max Quanchi explains in an essay about Hardy, "The text in *The Savage South Seas* was based on Hardy's reminiscences of his travels, written second-hand by Elkington" (2014: 219).
13. Sue Hodson, manuscript curator at the Huntington Library, confirms in an email of January 11, 2016, that the twelve thousand photographs in the Jack London collection contain no photographs of skulls, skull shrines, or canoe houses with skulls in them.
14. In a letter to Edward C. Marsh, an editor at Macmillan, London wrote to complain about placement of the illustrations for *The Cruise of the Snark*: "I gave the title to each photograph, the chapter in which such photograph was to appear, and the order in which it was to appear in such chapter (26 February 1911; London 1988: 2: 982). In a subsequent letter to Harold S. Latham, he complained about the "wreckage of the illustrations" in which some captions apparently had become separated from the intended photographs (7 April 1911; London 1988: 2: 997–998).
15. David Farrier makes a similar point about London's navigation of danger in the South Seas: "The settings through which he moves are themselves granted validity according to his capacity to occupy and dominate: there would be no 'there' if London were not there himself. Presence, there, in London's work, is more closely aligned with a force of will, with the definition of self" (2007: 179).

References

Arens, William. 1980. *The Man-Eating Myth: Anthropology and Anthrophagy*. New York: Oxford University Press.
Aswani, Shankar. 2000. "Changing Identities: The Ethnohistory of Roviana Predatory Head-Hunting." *Journal of the Polynesian Society* 109: 39–70.
Bennett, Judith A. 1987. *Wealth of the Solomons: A History of a Pacific Archipelago, 1800–1978*. Honolulu: University of Hawaii Press.
Castile, Philip. 2012. "The Last Phase of the South Sea Slave Trade: Jack London's *Adventure*." *Pacific Studies* 35: 325–341.
Cayley-Webster, Herbert. 1898. *Through New Guinea and the Cannibal Countries*. London: T. Fisher Unwin.
Cooper, Matthew. 1979. "On the Beginnings of Colonialism in Melanesia." In *The Pacification of Melanesia*, ed. Margaret Rodham and Matthew Cooper, 25–42. Lanham, MD: University Press of America.
Docker, Edward W. 1970. *The Blackbirders: The Recruiting of South Seas Labor for Queensland, 1863–1907*. Sydney: Angus & Robertson.
Dureau, Christine. 2000. "Skulls, *Mana* and Causality." *Journal of the Polynesian Society* 109: 71–97.
Elkington, Ernest Way. 1907. *The Savage South Seas*. Painted by Norman H. Hardy. Described by E. Way Elkington. London: A. & C. Black.
Farrier, David. 2007. *Unsettled Narratives: The Pacific Writings of Ellis, Stevenson, Melville, and London*. New York: Routledge.
Grimshaw, Beatrice. 1908. "In the Savage South Seas." *National Geographic* 19 (January): 1–19.
Imperato, Pascal James, and Eleanor M. Imperato. 1992. *They Married Adventure: The Wandering Lives of Martin and Osa Johnson*. New Brunswick, NJ: Rutgers University Press.
Jack London Papers, Huntington Library, San Marino, California.
Jahoda, Gustav. 1999. *Images of Savages: Ancient Roots of Modern Prejudice in Western Culture*. London and New York: Routledge.
Johnson, Martin. 1913. *Through the South Seas with Jack London*. New York: Dodd, Mead.
Johnson, Osa. 1940. *I Married Adventure: The Lives and Adventures of Martin and Osa Johnson*. Philadelphia: Lippincott.

Labor, Earle. 1962. "Jack London's Symbolic Wilderness: Four Versions." *Nineteenth-Century Fiction* 17: 149–161.
London, Charmian. 1915. *The Log of the Snark.* New York: Macmillan.
London, Jack. 1911a. *Adventure.* New York: Macmillan.
London, Jack. 1911b. *The Cruise of the Snark.* New York: Macmillan.
London, Jack. 1988. *The Letters of Jack London.* Ed. Earle Labor, Robert C. Leitz III, and I. Milo Shepard. 3 vols. Stanford, CA: Stanford University Press.
London, Jack. 2008. "Revolution." In *The Radical Jack London: Writings on War and Revolution,* ed. Jonah Raskin, 139–156. Berkeley: University of California Press.
London, Jack. n.d. "*Snark* Invoices." Jack and Charmian London Correspondence and Papers, 1894–1953 (COLL MSS 10), Box 30, Folder 10, Special Collections and Archives, Utah State University.
Lyons, Paul. 1995. "From Man-Eaters to Spam-Eaters: Literary Tourism and the Discourse of Cannibalism from Herman Melville to Paul Theroux." *Arizona Quarterly* 52 (2): 33–62.
Obeyesekere, Gananath. 2005. *Cannibal Talk: The Man-Eating Myth and Human Sacrifice in the South Seas.* Berkeley: University of California Press.
O'Brien, Aoife. 2017. "Crime and Retribution in the Western Solomon Islands: Punitive Raids, Material Culture, and the Arthur Mahaffy Collection, 1898–1904." *Journal of Colonialism and Colonial History* 18 (1): n.p. Project MUSE, doi:10.1353/cch.2017.0000.
Quanchi, Max. 2014. "Norman H. Hardy: Book Illustrator and Artist." *Journal of Pacific History* 49: 214–233.
Reesman, Jeanne Campbell. 2009. *Jack London's Racial Lives: A Critical Biography.* Athens: University of Georgia Press.
Sontag, Susan. 1973. *On Photography.* New York: Farrar, Straus and Giroux.
Thomas, Nicholas. 1989. "The Force of Ethnology: Origins and Significance of the Melanesia/Polynesia Division." *Current Anthropology* 30: 27–41.
Tucker, Amy. 2017. "Charmian's 'One True' *Log of the Snark.*" *Women's Studies* 46: 362–387.
Waite, Deborah. 2000. "An Artefact/Image Text of Head-Hunting Motifs." *Journal of the Polynesian Society* 109: 115–144.
Woodford, Charles M. 1890. *A Naturalist Among the Head-Hunters.* London: George Philip.
Wright, Christopher. 2013. *The Echo of Things: The Lives of Photographs in the Solomon Islands.* Durham, NC: Duke University Press.

PART III

CREATING AND RECOVERING PERSPECTIVE

Chapter 8
Forgetting London
Paris, Cultural Cartography, and Late Victorian Decadence
Alex Murray

In *Confessions of a Young Man* George Moore stated: "To write about London I should have to begin by forgetting Paris." In what follows I would like to interrogate this premise that immersion in a foreign culture can have a drastic effect on the subjective experience of place. For decadent and aestheticist writers such as Moore and Arthur Symons Continental travel and cultural exchange represented a paradigmatic shift in their experience and understanding of themselves as British subjects.[1] This shift is clearly evident in the antagonistic relationship between London and Paris that exists for both writers. For Moore and Symons the experience of Paris would inevitably come to an end. For Symons it lasted little more than three months before he returned to London. Moore, after a residence of seven years, was forced to return to Ireland to oversee his estates. Yet the experience of Paris was to have a far more profound effect than recollection of youthful folly. I would like to argue that both writers on returning from the Continent attempted to re-create what they perceived as the radically avant-garde literary practices of Paris. In this re-creation they began to refract London, to significantly alter the ways in which the city was represented, and in doing so alter forever the way in which this modern metropolis was represented.

The Victorian Cosmopolis

One of the most seductive critical paradigms for imagining Victorian London is as a cosmopolis, the center for a literary and cultural production that was increasingly global, the hub of a networked world of literary cultures. The recent "cosmopolitan" turn in Victorian studies has done much to radicalise the Decadent movement by rejecting its charges of political quietism through the identification of a decadent form of "community." The argument here is

---

Notes for this chapter can be found on page 148.

that the rejection of forms of nationalism, and the embracing of small international communities of shared "taste" constitutes a radical idea of community. As Matthew Potolsky has argued "Against the organically unified nationalist canon and the national character it ostensibly reflects, decadent writers posit a transnational and transhistorical 'decadent subject' as an alternative to the nineteenth-century ideals of national community." (2006: 216) Elsewhere Potolsky has referred to this as a "decadent counterpublic," emphasizing the sense of a radical alternative to the dominant public sphere of the nation (2007).

Yet for all the political possibilities of radical community, Potolsky and others fail to see the paradox that reducing Decadence to the status of a "counter" public still leaves the very logic of the "public" and of national community intact. To my mind the idea of "decadent cosmopolitanism" in this guise represents a dangerous flirtation with the idea of an insular and elitist decadence that leads to the ossified institutions and coteries of modernism. At one point Potolsky refers to Jean-Luc Nancy's notion of the "literary communism," approvingly stating "The decadent community 'finds' itself, perhaps, in an artificial and temporary assemblage of individuals, much as Des Esseintes and Dorian Gray 'find' themselves in their collections." (2006: 241). Like Nancy's community, decadence is characterised by a "union of artificial singularities" (241) that refuses subsumption into a whole. Yet this seems to be a limited reading of Nancy's idea of "community," with Potolsky ignoring Nancy's notion of an "inoperative community" or a *communite désoeuvre* which, to my mind, gives us a far more potent "political" decadence. Instead of falling into counterpublics, which are after all a smoke screen for neoliberalism (like our contemporary communities of difference), our study of decadence must return to questions of linguistics and form, to uncover a truly "inoperative" form of community based upon a deactivation of the very linguistic essentialism that underpinned Victorian ideas of national identity.

Perhaps we could turn here to the idea of the "inoperative community" that Nancy advocates and Potolsky overlooks. The French term for "inoperativity," *désœuvrement*, is important here, and a sensitivity to its complex registers in post-war French thought can unpick the importance of "communication" in relation to 'community'. Its use in postwar philosophy stretches back to Alexandre Kojève and Georges Bataille, who debated its meaning in the 1950s. It was then used by the French novelist and philosopher Maurice Blanchot. As we can see clearly it is an inversion of the word "œuvre" or body of work, so it would seem to be a non-work, or an unworking, inertia, lack of work. Bruce Baugh, in *French Hegel* uses the term "poetic undoing" to translate the term from Bataille's essay on surrealism, and I think it is important not to lose sight of the properly literary form that *désœuvrement* and inoperativity take

(Baugh, 2003: 76) Yet there is also, for Jean-Luc Nancy, a properly "political" form of *désœuvrement*. As Nancy states:

> the community takes place of necessity in what Blanchot has called *désœuvrement*... the community is made up of the interruption of the singularities, or of the suspension singular beings are. It is not their work, and it does not have them as it works, not anymore than communication is a work... Communication is the unworking of the social, economic, technical, institutional work (Nancy, as in Joris, 1988: xiv).

It is the idea of communication and communicability as the unworking of the social and linguistic institutions of work through the act of suspension, or what I will call, following Agamben, deactivation, that can lead us to an understanding of the challenge decadence provides to the nation state. In doing so I intend to move the study of decadence away from a limited idea of politics based on some form of collectivity, and instead return to its properly political form, the poetic.

Travel, Decadence, London: Toward a Cultural Cartography of the *Fin de Siècle*

In order to explore the political nature of decadent form in relation to national identity it is necessary to outline, albeit briefly, the ways in which Continental travel contributed to the decadent challenge to national identity. The second half of the nineteenth century has long been seen as a pivotal point in the development of modernity. While the industrial revolution of the eighteenth and early nineteenth centuries had unparalleled effects on many lives, the late Victorian period witnessed an acceleration in the development of transport and communication that profoundly affected urban sensibilities. In particular there was a rapid increase in the speed and affordability of long-distance travel. No longer was European touring restricted to the aristocratic and upper-middle class; instead it had become democratized, available to an upwardly mobile bourgeoisie who were desperate to accumulate the cultural capital provided by travel. Perhaps we can take as indicative of this development the Meagles family in Dickens's *Little Dorrit*. The family is certainly not overly wealthy, yet they desire the acculturation of the upper classes. Their experience is marked as largely touristic, concerned with the accumulation of cultural artefacts as synecdoches of experience that can be displayed in the bourgeois home. Arthur Clennams's first visit to the Meagles home in Twickenham reveals this desire to posses the material traces of travel rather than the experience itself: "Of articles collected on his various expeditions, there was such a vast miscellany that it was like the dwelling of an amiable corsair." Then follows a taxonomy of objets d'art collected by Mr. Meagles in his travels, before they move on to his "pictorial acquisitions." "He was no judge, he said, except of what pleased himself; he had picked them up, dirt-cheap, and people *had* considered them rather fine" (1982: 163). Here it is the cultural authority of others that bestows value, a recognition of being cultured, rather than what we might think of as a quest for authenticity and experience that marked such "travel" experiences of many of the famous Romantic writers. The democratization of travel and the rise of tourism thus represents a shift in the forms of European travel. With these changes comes an ambivalence over the place of cultural exchange.

James Buzard's influential study *The Beaten Track: European Tourism, Literature and the Ways to 'Culture'* (1993) has been of great value in our understanding of the complex systems of cultural value that circulated around travel and tourism in the Victorian period. As Buzard asserts:

> Modern tourism has tended to reinforce existing privileges, reproducing assumptions about the special suitedness of well-to-do northern European men for fully realised acculturation. Licensing the notion of a superior sensibility that apparently owes nothing to social conditions, it has in effect masked differences in degree of freedom from economic necessity and in what Pierre Bourdieu calls 'habitus', the internalized system of 'dispositions' which, among things, prepare one for the satisfactory appropriation of cultural goods (1993: 7).

For Buzard the rise of tourism must be seen in relation to its other, namely travel, the quest for an authentic experience of place, the adventure off the beaten track. It is essential that we begin to identify and explore the "dispositions" which underpin travel and to understand the desires that circulate around the accumulation of travel. Yet I would argue that we can perhaps extend this analysis of travel by exploring precisely what the acculturated male travelers of the avant-garde brought home with them. I would like to suggest that for writers such as George Moore and Arthur Symons, travel was not simply about the accumulation of cultural capital and objects d'art. Instead, travel was about exploring and then emulating certain cultural practices which were radically at odds with the dominant paradigms of national identity. Buzard suggests that instead of looking to the experiences of the tourist and traveler as unmediated experience of the foreign, we should look at them as reflected experience of home: "the dichotomy (traveller/tourist) has more to do with the society and culture that produce the tourist than it does with the encounter any given tourist or 'traveller' may have of a foreign society and culture" (1993: 5). I would agree that this is the case, but would like to suggest that it doesn't necessarily have to be all one-way traffic. The desires—in many ways elitist –of the decadent writers of the late-Victorian period to shatter the hegemony of English culture were certainly projected onto sites and locales in Europe, yet on their return the experience of foreign culture can be seen in a transformation of English locales—in this case London—into an unfamiliar and uncanny space.

As this essay seeks to complicate our understanding of late-Victorian travel, it also attempts, as I suggested above, to complicate and extend the discourse surrounding decadence. Decadence is often regarded as a transgressive, almost apolitical form of rebellion from the high-moralism of the Victorians. This reading, focusing on Oscar Wilde, seeks to celebrate decadence as an important step in the development of homosexual identity. The specific politics of decadence can become lost as we begin to teleologically trace lineages to a specific debate issue within our own political and cultural climate, potentially obfuscating the broader cultural context of decadence. In exploring late decadence and what I will term cultural cartography, I am suggesting that decadence can be recast as an anti-nationalist literary movement, one that sought to reject the equivalence between place and identity that was such an important feature of Victorian identity. Significantly, it does so not through the establishment of transnational literary coteries but by attacking the very representation of place.

In thinking of the Victorian period and national identity, Matthew Arnold often emerges as a key figure in the nexus between the nation, culture, and Europe. Arnold's concern over the cultural health of England was intrinsically linked to the relationship between foreign and domestic culture, a relationship that permeated the work of both the proponents and critics of decadence. Arnold believed that allowing foreign discourse to subtly penetrate English thought was a vital means of cultural regeneration. In his 1865 essay "The Function of Criticism at the Present Time," Arnold asserts:

> An epoch of expansion seems to be opening in this country. In the first place all danger of a hostile forcible pressure of foreign ideas upon our practice has long disappeared; like the traveller in the fable, therefore, we begin to wear our cloak a little more loosely. Then, with a long peace, the ideas of Europe steal gradually and amicably in, and mingle, though in infinitesimally small quantities at a time, with our own notions. (1911: 17).

Arnold's most famous representation of such an idea is his poem "Dover Beach." What we see in "Dover Beach" is the paradoxical obsession with cultural regeneration and maintaining a national culture. In the poem there is certainly a sense that the stability of English identity, symbolized in the cliffs of Dover, retains an ambivalent relationship to the setting sun on the Continent. Does the light suggest, as Julian Wolfreys has asserted, the ephemerality of French thought (1994: 32), or could it be a form of cultural illumination that could avoid the clashes of England's "ignorant armies"? The ambivalence here is key to a central Victorian anxiety about cultural transmission and exchange. For the Decadent writers who followed Arnold, the light of French thought constituted an invigoration that could revise a dormant English language and culture. Yet simultaneously they saw in it the undermining of the conservatism that was antithetical to cultural rejuvenation.

It is this anxiety, I would like to suggest, that a number of key decadent writers were able to explore and exploit through their engagement with foreign culture. While much has been written of the circulation of literary ideas across the Channel, I would like to look instead at physical movement across the Channel. There can be a tendency to over-textualize decadence, looking for clues to its development in the work of Huysmans, Baudelaire, Mallarmé, Gautier, and Verlaine. While these writers are significant, I argue that we also need to look at the ways in which writers experienced culture and cultural spaces. The journey to Paris, made by close to every decadent writer of the 1890s, was essential in the romance of Continental culture that was so essential to decadence. It was not simply the literary discoveries, but new ways of experiencing place, new forms of affective mapping of the city that was to prove so vital. These ideas, these experiences of space were then transported to London, challenging the Victorian equivalence between the metropolis and the empire that were so crucial to maintaining its cultural hegemony.

George Moore, or Parisian Naturalism in London

And it is to Paris that this article will turn to begin the process of mapping the ways in which the cultural cartography of Europe began to alter the literary cartography of London. I would argue that it is this process of cultural exchange, of altering the representation of space through the experience of place, that can help us develop our understanding of this causal relationship. George Moore, born into wealthy Irish gentry, arrived in Paris in March 1873. His first impressions of the city were underwhelming, to say the least:

> We all know the great grey and melancholy Gare du Nord at half-past six in the morning; and the miserable carriages, and the tall, haggard city. Pale, sloppy, yellow houses; an oppressive absence of colour; a peculiar bleakness in the streets... a dreadful *garçon de café*, with a napkin tied round his throat, moves about some chairs, so decrepit and so solitary that it

seems impossible to imagine a human being sitting there. Where are the Boulevards? Where are the Champs Elysées? I asked myself, and feeling bound to apologize for the appearance of the city, I explained to my valet that we were passing through some side streets (1925: 12).

In this passage Moore is deliberately establishing his initial experience of the city as that of the tourist. Using the universal "we," he attempts to posit this impression as one that all visitors to Paris experience: the sense of disappointment and disdain stem from the city's inability to fulfill its own place in the cultural imaginary of the polite tourist. For a tourist, Paris is constituted of sites and locations of culture, and for the pre-Parisian Moore, their absence can only be regarded as an inadequacy. His need to apologize is the need to affirm a sense of the touristic impulse to consume the foreign in pre-imagined forms, in ways that pose no threat to the national identity of the tourist.

For Moore the shift from the tourist to the intrepid traveler was one that was linked to cultural immersion, to a refashioning of the self at the hands of a foreign culture. The art studios in which he undertook instruction, the cafés and literary salons were to be his sites of education. As he asserted in *The Confessions*: "I did not go to either Oxford or Cambridge, but I went to the 'Nouvelle Athens'" (1925: 85). In this idea of an alternative pedagogy, Moore was framing his time in Paris as an education, one that was to challenge many of the ideals that had been instilled in him during his privileged childhood. His description of his re-education is littered with literary references, and indeed Moore became an expert in contemporary French literature as he consumed the works of Verlaine, Baudelaire, Rimbaud, and Zola. This immersion in French poetic practice had a dual effect on the way in which Moore responded to the literature of his native tongue. Firstly it made him aware of certain inadequacies. As he said of a Verlaine sonnet: "no English sonnet lingers in the ear like this one, and its beauty is as inexhaustible as a Greek marble" (1925: 66). This statement represents both a denunciation of English poetic practice, and also, through the rather obsequious reference to Pater, a commentary on the lack of aesthetic autonomy among his own countrymen (see Pater, 1972: 77-8). The second central effect of Moore's immersion in French literature was the loss of his native tongue. As Moore asserts: "I have heard of writing and speaking two languages equally well, but if I had remained two more years in France I should never have been able to identify my thoughts with the language I am now writing in, and I should have written it as an alien" (1925: 124). While this experience of language represents a fascinating insight into the relationship between language and national identity, I would instead like to focus on how Moore's experience of Paris resulted in his creation of a new model of representation that was to later provide such a *succès de scandale* when transported to England.

Moore, as always, represents experience as self-fashioning, as what Grubgeld has called the autogenic conception of the self (1994). In this conscious self-fashioning, Moore came to develop a markedly different conceptualization of not only himself, but his perception of the world around him. Heavily influenced by Zola and Manet, Moore used his experience of Paris to develop a scientific model of observation and representation that, through its naturalistic imperative, was to challenge the moralizing narratives of Victorian literature. In *The Confessions* we see this shift in perspective as Moore moves from the idealistic young man who was repulsed by Paris on first arrival to the detached and critical Zolaesque observer. Describing visiting a wealthy salon he states:

> just as I had watched the chorus girls and mummers, three years ago, at the Globe Theatre, now, excited by a nervous curiosity, I watched this world of Parisian adventurers and lights-o'-love. And this craving for observations of manners, this instinct for the rapid notations of gestures and words that epitomize a state of feeling, of attitudes that mirror forth the soul, declared itself a main passion (1925: 24).

Here passion becomes refracted through the detached observation of Moore's peculiar naturalism, desire is transformed into an obsessive, descriptive gaze.

It was this passion that Moore brought back to London in 1880 as he tried to re- invent himself as a novelist and man of letters. The passion was also to manifest itself in an altering representation of the city, as the imaginative manifestation of reality in Dickens gave way to an immoral science of writing, one that was to reduce London to no more than the site of a series of desires, both sexual and social. Moore's notorious 1883 novel, *A Modern Lover*, encapsulates this transformation of the city through representation. For a writer such as Moore, modern life wasn't constituted by omniscient and timeless values, but by transitory and irrational human desire. This form of desire is one that Moore attempts to dramatize in his autobiographical writings, as a desire to create oneself overshadows any notion of qualities. As he states in the opening page of *Confessions*: "I may say that I am free from original qualities, defects, tastes, etc … I came into this world apparently with a nature like a smooth sheet of wax, bearing no impress, but capable of receiving any" (1925: 1). This notion of being a parasitic individual with nothing as stable as an essence suggests that identity is constituted of nothing more than manifold experiences, amounting to little more than detached wonder at the transience of identity. Indeed for Moore, the act of writing is one that is designed to both create and obliterate, construct and deconstruct the self. This process is analogous to what Paul de Man describes as the de-facement of the autobiographical: "Death is a displaced name for a linguistic predicament, and the restoration of mortality by autobiography (the prosopopeia of the voice and the name) deprives and disfigures to the precise extent it restores. Autobiography veils a de-facement of the mind which it is itself the cause" (1984: 81). This sense of writing as deconstructive imbues both Moore's numerous autobiographical texts, as well as his novels. In both works, character is consistently reworked to the point where it is a sheet of wax, or, to use de Man's terminology, the thinnest veil on the surface of the non-self.

In *A Modern Lover*, a sense of detached wonder is presented through the narrative perspective which attempts to challenge and undermine any sense that the city is the site of imaginative response. The central character, Lewis, a destitute bohemian artist, wanders the city in search of experience. In the opening chapter he emerges into the Strand as a crowd streams out of a theater. He then proceeds to describe a scene, but in describing the scene he seems to describe himself out of existence: "Lewis listened, and soon losing sight of his own personality, saw the scene as an independent observer, and dreamed of a picture to be called 'suicide'" (1883: 13). As the role of observation destroys subjectivity, so does Moore's clinical narrative tone, as it sets about to destroy the illusion of London as anything more than a site which the imagination seeks to obfuscate. Lewis's experience is always represented as one that seeks to project his own turmoil onto the blank slate of the city. Take for instance Lewis gazing upon the Thames after leaving the overwhelming color and movement of the Strand:

> All was fantastically unreal, all seemed symbolical of something that was not. Along the embankment, turning in a half circle, the electric lights beamed like great silver moons, behind which, scattered in inextricable confusion, the thousand gaslights burned softly like night-lights in some gigantic dormitory... The mystery of the dark wandering waters suggested peace, and in the solemn silence he longed for the beatitude that death can only give, as in the glitter and turmoil of the Strand he had yearned for the pleasures of living (1883: 14-15).

There are several important elements in this passage. Firstly, the location of this passage is far from arbitrary. The passage, which takes place on Waterloo Bridge, mirrors Wordsworth's poem "Lines composed upon Westminster Bridge." In this poem, the great Romantic poet transforms the city he was later to critique so stridently in the "Prelude" into an imaginative reflection of his youthful enthusiasm upon journeying to France. In response, Moore's Lewis attempts to highlight the transitory and insincere nature of imaginative projection. The fantastical unreality is transported into prose as language struggles to represent this experience. The projected imagination of the artistic visionary doesn't transform the scene; instead that is left to the electrified lights as Lewis becomes aware that his attempts to imbue the city with a metaphoric gravity fail. This undermining of the position of the artist then opens up the realism of the novel to greater effect. The observations of human activity aren't to be compromised by an imaginative conceptualization of the city that would give it some sort of Dickensian quality as a paradigmatic force in the life of its characters.

Razing Paris: Impressionism and the Negation of Place

Moore's experience is paralleled by that of Arthur Symons, who in 1890, accompanied by Havelock Ellis, left London to spend three months in the French capital. Here the two young men indulged in the culture of the capital, attending the theater, Easter services at Notre Dame, and a lecture by Moore's mentor Catulle Mendès, as well as meeting Odilon Redon. Yet for Symons, this touristic immersion in culture gave way to something far more profound. On 25 May he wrote to J. Dykes Campbell, exclaiming:

> I am by this time getting so Parisian that the thought of London fills me with horror. I am contemplating permanent residence here; have forgotten most of my English (though I can still write it fairly well) and have begun to write in French for the 'public prints'. (Symons, in Beckson, 1987: 54)

This loss of identity through the loss of language mirrors that of Moore twenty years earlier. Paris had begun to overwhelm Symons's imagination, and in losing language he could, conceivably be losing a central element of his national identity. Indeed this sense of Paris, and of Parisian culture, as being antithetical to England was to be the focus of his discussion of the city some fourteen years later. In the indicatively entitled *Colour Studies in Paris*, Symons attempts to re-create the city in the evocative style of French impressionism. A collection of essays, poems, and impressions written between 1890 and 1907, the collection moves from Verlaine to Dancing, and from the gingerbread fair at Vincennes to Victor Hugo, neatly encapsulating Symons's dual obsessions: the high-cultural space of avant-garde literature and the popular cultural spaces of the music hall and the street. In an early poem featured in the collection entitled "Paris," originally penned in 1894, we see the ways in which Symons attempts to evoke the city via impression rather than identifying physical sites or cultural manifestations:

> My Paris is a land where twilight days
> Merge into violent nights of black and gold;
> Where, it may be, the flower of dawn is cold:
> Ah, but the gold nights and the scented ways!
>
> Eyelids of women, little curls of hair,
> A little red nose curved softly, like a shell,
> A red mouth like a wound, a mocking veil:
> Phantoms, before the dawn, how phantom-fair!
>
> And every woman with beseeching eyes,
> Or with enticing eyes, or amorous,
> Offers herself, a rose, and craves of us
> A rose's place among our memories
> (1918: 3)

Here we see that the proper place name of Paris is already negated by the possessive, subjective voice of the poet, "My," that singular, fluid and unstable signifier that will dictate the rest of the poem. Indeed "my" works to undo Paris, as the rest of the poem presents us with a series of ambivalent experiences and sensations that could be experienced in any number of places. In fact, as I will suggest a little later, they could even occur in London, and for Symons the importance here is not the place, but the poetic experience. Indeed the poem consists of a series of symbols—the rose, the red mouth, curls of hair, amorous eyes—that are designed to give the impression of female sexuality, suggesting that sexual freedom can be found in Paris, but certainly not to conjure up the city itself.

If the locale of the poem is to be of little importance for its content, then we can begin to see that what the experience of Paris has done is not simply create an obsession with French culture but transform the model of representation into one that denies the confluence between space and identity. This is made clearer in Symons's essay "Montmartre and the Latin Quarter," in which Symons describes these locales as "the two parts of Paris which are unique, the equivalent of which you will search for in vain elsewhere" (1918: 24). The singularity of these sites may seem antithetical to the arguments I have made for the experience of Paris of both Moore and Symons thus far, and indeed the essay here seems to suggest that it is because these sites and spaces are so antithetical to London that they are so singular to Paris: "There will never be a Boul' Mich' in London. It is as impossible as Marcelle and Suzanne. The Boul' Mich' is simply the effervescence of irrepressible youth; and youth in London never effervesces" (1918: 25). Yet this rather simple opposition between the two cities is certainly not at the heart of Symons's essay. Indeed I would argue that the essay sets up this binary only to then place upon Paris the ambivalent and fleeting impressionism that will subvert any spatial landscape in the city as being "essential."

The conclusion of Symons's essay coincides with the end of an omnibus journey near the top of Montmartre. As the omnibus makes its way north from the Odéon, Symons describes in fleeting detail passing certain sites and spaces: the Boulevard Saint-Germain, Saint Sulpice, the Louvre, and the rue de Rivoli. This cataloging of places and spaces would seem to give to Paris a concrete certainty, an equivocal link between space and the experience of the city. Yet upon climbing Montmartre, Symons seeks to call into question the sense of the city as a knowable series of spaces:

> Under a wild sky, as I like to see it, the city floats away endlessly, a vague, immense vision of forests of houses, softened by fringes of actual forest; here and there a dome, a tower, brings suddenly before the eyes a definite locality; but for the most part it is but a succession of light and shade, here tall white houses coming up out of a pit of shadow, there an unintelligible mass of darkness, sheared through by the inexplicable arrow of light. Right down below, one looks straight into the lighted windows, distinguishing the outline of the lamp on the table, of the figure which moves about the room, while in the far distance there is nothing but a faint, reddish haze, rising dubiously into the night, as if the lusts of Paris smoked to the skies (1918: 33).

The view from Montmartre is famously the clearest and most iconic view of the city as a whole, yet tellingly Symons celebrated the negation of the city, reducing it to the impression of light and dark, inside and outside, metaphorically raising the city in an impressionistic conflagration. To reduce the city to poetic impression is, necessarily, to shift and alter its relationship to the hegemonic forms of power and culture, to celebrate it instead as an indistinct and fluid zone of experience. Paris is not a series of spaces but the by-word for a particular form of impressionism.

Mapping the Decadent Capital: Toward a Disappearing London

In what follows I will suggest the ways in which Symons's impressionistic articulation of urban life reveals the influence of European travel in the representations of that most British of spaces. Upon returning from Paris in 1891, Symons brought back with him the aesthetic of Verlaine, Huysmans, and Mallarmé. Symons found Verlaine's poetry "as lyrical as Shelley's, as fluid, as magical—though the magic is a new one. It is a twilight art, full of reticence, of perfumed shadows, of hushed melodies. It suggests, it gives impressions, with a subtle avoidance of any too definite or precise effect of line or colour... The impressions are remote and fleeting as a melody evoked from the piano by a frail hand in the darkness of a scented room" (1918: 171-2). It is this attempt to capture the transient and subtle nature of the impression that marks Symons's poetry of the 1890s, along with an attempt at capturing the *recherché* of the musical hall and London prostitution. His 1892 volume of poetry *Silhouettes*, and 1895 volume *London Nights* attempt to emulate this form of evocative and impressionistic French verse. For Symons, it was the ability of impressionism to record the experience of the city that made it the only form of truly modern poetry. As he stated in his essay "Modernity in Verse" when reflecting on the poetry of William Ernest Henley, "I think that might be the test of poetry which professes to be modern: its capacity for dealing with London, with what one sees or might see there, indoors or out" (1897: 188). This idea of a poetry that could capture London was one that shifts the experience of the city from archetype to impression, at the same time representing a London that began to lose its specificity.

London Nights and *Silhouettes* are littered with this attempt to capture London anew. One indicative example is the poem "In Fountain Court":

The fountain murmuring of sleep,
    A drowsy tune;
The flickering green of leaves that keep
    The light of June;
Peace, through a stumbling afternoon,
    The peace of June
A waiting ghost, in the blue sky,
    The white curved moon;
June hushed and breathless, waits, and I
    Wait too, with June;
Come through the lingering afternoon,
    Soon love, come soon
(1892: 85)

The title of the poem refers to Symons's lodgings in the period he wrote this poem, the well-known Fountain Court near the Strand. Yet the specificity of the title is absent from the poem. Here Fountain Court cannot be represented by its proximity to Temple Bar or the architecture of the surrounding buildings or the thriving intellectual community that resided there in the 1890s. Instead, it is the impression of a June afternoon, the murmuring fountain, the play of light that has replaced urban space. The personal recollection, the play of memory as a series of impressions, overwhelms the poem's site-specific object. As in a great deal of Symons's poetry, even the content begins to fade away as the lilting repetition of sounds forms its own impression, with the locality of the poem becoming twice removed. It was this absence of a tangible London that so frustrated Symons's contemporary, the Decadent Catholic poet Lionel Johnson. As Yeats recalls: "Arthur Symons' poetry made him angry, because it would substitute for that achievement [of the intellect] Parisian impressionism, 'a London fog, the blurred tawny lamplight, the red omnibus, the dreary rain, the depressing mud, the glaring gin-shop, the slatternly shivering women, three dexterous stanzas telling you that and nothing more'" (Johnson in Yeats: 237-8) As Johnson correctly identifies, this is a poetic form that is distinctly foreign to the city, and, as such, I would argue it is partially responsible for changing the artistic representation of the city.

What we begin to see as we investigate Symons's model of representing the city is the substitution of locality for not just the fleeting impression, but its inclusion as linguistic signifier with no reference to any tangible reality. This notion of locality being excluded through its inclusion as object is captured perfectly in the poem "Impression." Here London is evoked not to locate the action of the poem, which is not existent, but as a word with its syllabic tone included for the sake of a rhythmic impression:

> Outside, the dreary church bell tolled,
> The London Sunday faded slow;
> Ah what is this? What wings unfold
> In this miraculous rose of gold?
> (1892: 15)

The staccato rhythm of this second line is used to give further effect to the rhetorical flourish that closes the poem, to leave the language all the more urgent and pressing. This use of language for rhythmic purposes is a key feature of the decadent desire to display that autonomous nature of language by fracturing signifiers from referents. Here London is not a place, a metropolis, a site of empire, a paragon of civilization, but nothing more than a word that can help produce the rhythmical life of language that would free it from the onerous imperative for meaning.

If London was invoked in *Silhouettes* in order to fracture it from representation, in *London Nights* the city is paradoxically absent. Here Symons uses the city in the title to precisely undermine the notion that a city is constituted of anything more than the manifold impressions and thoughts of those who live there. London is simply a space in which the properly universal accumulation of symbols and impressions can take place. London here loses its own identity, subsumed into Symons's own idiosyncratic experience of the city. Take one of Symons's most famous poems, "Nora on the Pavement." It is a brief yet complex composition that attempts to represent the freedom of urban life through the consonant and dissonant nature of rhythm.

> As Nora on the pavement
> Dances, and she entrances the grey hour
> Into the laughing circle of her power,
> The magic circle of her glances,
> As Nora dances on the midnight pavement;
>
> Petulant and bewildered,
> Thronging desires and longing looks recur
> And memorably re-incarnate her,
> As I remember that old longing,
> A footlight fancy, petulant and bewildered;

Here the poem uses a rather forced and jarring rhythm with an excess of syllables to create the effect of claustrophobia. This rhythm could be said to evoke that of the industrialized automated impression of life in the city, in which the ceaseless rhythm constrains and contorts poetic control. This is contrasted to the final lines of the poem in which Nora and the rhythm are set free:

> Herself at last, leaps free the very Nora
>
> It is the soul of Nora,
> Living at last, and giving forth to the night,
> Bird-like, the burden of its own delight,
> All its desire, and all the joy of living,
> In that blithe madness of the soul of Nora.
> (1895: 7)

Here Nora's freedom is encapsulated with a far more lilting cadence, suggesting that Symons's poetic practice both takes its rhythm from the transcendent soul of Nora and creates the representation that frees Nora. Here the poetic rhythm, in transcending the constricting rhythms of modern life, is portrayed as the antidote to London, with Symons's poetic practice superimposing the imaginary of the poet over the reality of the city. This effect of imposing a model of representation over the city in order to alter what was perceived as constituting its reality led one early critic, Stange, to assert that Symons "becomes increasingly able to give the scattered facts of the city a metaphorical significance and is, in fact, the first English poet who was able to write about London with something like Baudelaire's mythographic sense, to make the city a convincing milieu of spiritual adventures." (491)

If we reflect on the massive influence that Symons had on those poets of the modernist city such as Eliot and Pound, his writing, in moving the city from symbol of empire to the site of manifold impressions, has had a lasting legacy in representations of the city. Similarly Moore's attempt to shatter both the Romantic process of subjectivization and the archetypal model of the city as dramatic entity is one that became common in twentieth- century representations of the city. Yet to give either of these writers a paradigmatic position in the shifting representations of London in literature would be misleading. Instead, I would argue that we can see these writers as demonstrative of a host of influences that have shaped the city and its representation. For Symons and Moore it was the experience of Paris that proved dramatic in their perception of the city. Their attempt to map London was one whose cultural cartography has its foundations, paradoxically, in the streets of Paris. Here the immersion in French culture—and in particular the decadent literature emerging from the French capital—was transported to London, transforming the city in the process.

That transformation of the city, its representation as an "other" city—Paris—needs to be read as a specifically political transformation. London, that center of empire, becomes deactivated here through the importation of a Parisian impressionism. That is not to say that London becomes Paris, but that London is shed of its associations with the city of empire, losing its specificity to emerge as a site of myriad impressions, a space in which the illusory imagined communities of the nation state disappear as they are replaced with forms of contingent, singular, and fleeting communication that end up communicating little more than communication itself.

Alex Murray is Senior Lecturer in Modern Literature at the Queen's University, Belfast. His most recent monograph is *Landscapes of Decadence: Literature and Place at the Fin de Siècle* (Cambridge University Press, 2016) and he has edited *Decadence: A Literary History* (Cambridge University Press, 2020) and, with Kate Hext, *Decadence in the Age of Modernism* (Johns Hopkins University Press, 2019).

Note

1. Even though Moore was Irish his self-identity was, in his early life at least, antithetical to any concept of a national identity. To refer to him as a "British" writer is more a gesture toward the relationship between language and national identity rather than any argument about his own sense of self.

References

Arnold, Matthew. [1865] 1911. *Essays in Criticism*. London: Macmillan.
Baugh, Bruce. 2003. *French Hegel: From Surrealism to Postmodernism*. London: Routledge.
Beckson, Karl. 1987. *Arthur Symons: A Life*. Oxford: Clarendon Press.
Buzard, James. 1993. *The Beaten Track: European Tourism, Literature and the Ways to Culture, 1800-1918*. Clarendon Press: Oxford.
de Man, Paul. 1984. *The Rhetoric of Romanticism*. New York: Columbia UP.
Dickens, Charles. 1982. *Little Dorrit*. Oxford: OUP.
Gagnier, Regenia. 2005. "Morris's Ethics, Cosmopolitanism, and Globalisation" *Journal of William Morris Studies*: Summer and Winter 2005: 9-30.
Grubgeld, Elizabeth. 1994. *George Moore and the Autogenous Self: The Autobiography and Fiction*. New York: Syracuse University Press.
Holdsworth, Roger. 1989. 'Introduction' to Arthur Symons, *Selected Writings*. Manchester: Carcanet: 9-24.
Joris, Pierre. 1988. 'Translator's Preface' Maurice Blanchot *The Unavowable Community*. Barrytown: Station Hill Press: xi-xxix
Moore, George. 1883. *A Modern Lover: A Realistic Novel*. London: Walter Scott.
——— [1888] 1926.*Confessions of a Young Man*. London: William Heinemann.
Pater, Walter. 1972. *Essays on Art and Literature*. Ed. Jennifer Uglow. London: J.M. Dent.
Potolsky, Matthew. 2006. "Decadence, Nationalism and the Logic of Canon Formation" *Modern Language Quarterly, 67: 2*: 213-244.
——— 2007. "The Decadent Counterpublic", *Journal of Romanticism and Victorianism on the Net 48*. http://www.erudit.org/revue/ravon/2007/v/n48/017444ar.html (accessed 4/19/2010).
Stange, G Robert. 1973. "The Frightened Poets." Pp. 475-94 in H.J. Dyos and Michael Wolff, eds. *The Victorian City*, Volume Two. London: Routledge, Keegan and Paul.
Symons Arthur. [1892] 1993. *Silhouettes*. Oxford: Woodstock Books,
——— [1895]. 1993. *London Nights*. Oxford: Woodstock Books.
——— 1897. *Studies in Two Literatures*. London: Leonard Smithers.
——— 1918. *Colour Studies in Paris*. London: Chapman and Hill.
Wolfreys, Julian. 1994. *Being English: Narratives, Idioms and Performances of National Identity from Coleridge to Trollope*. Albany: SUNY Press.
Yeats, William Butler. 1999. *The Complete Works:* Volume III, Autobiographies. New York: Scribner.
Young, Paul. 2009. *Globalization and the Great Exhibition: The Victorian New World Order*. Basingstoke: Palgrave.

Chapter 9

IN THE EYES OF SOME BRITONS
Aleppo, an Enlightenment City

Mohammad Sakhnini

[T]hat in Aleppo once
Where a malignant and a turban'd Turk
Beat a Venetian and traduc'd the state
I took by the throat the circumcised dog
And smote him—thus. [stabs himself]
—William Shakespeare, *Othello,* 1604

Eighteenth-century British representations of the diverse and multireligious Middle Eastern city Aleppo allow us to reimagine the narrow exclusionary frontiers of identity that recently emerged in contemporary reactions in Britain toward the idea of crossing national and cultural borders. The recent debate in Britain about refugees crossing physical borders from Syria and elsewhere revealed many people's sensitivities surrounding cultural and religious borders. And the recent referendum that resulted in Britain leaving the European Union, commonly known as Brexit, saw, in its lead-up, a platform warning of the danger of Turkey joining the European Union and the influx of Syrian refugees to the country. The leave campaign's rhetoric concerning the arrival of more Muslims in Britain touched many in the country as a scaremongering tactic, but there was also a sense of historical amnesia regarding Britain's long history of being in the world, particularly the longstanding relationships and interactions between Britain and the Middle East that extend beyond the narrow fear of Islam and Muslims (Laidlaw 2010; Mather 2011). This article thus aims to rethink this historical amnesia by focusing on the accounts of a few British travelers who visited and lived in Aleppo throughout the eighteenth century to show how recent British fear of Islam and Muslim refugees is not consistent with previous cosmopolitan encounters many Britons experienced in eighteenth-century Syria. These Britons, mostly from the middle and upper classes who visited the East as merchants and curious travelers, found in Aleppo a cosmopolitan space where Christians, Muslims, and Jews lived in

---

Notes for this chapter begin on page 165.

relative harmony and peace, a traveling experience that serves to highlight how their identities, both particularist and globalist, were steeped in the context of their entanglement with the world of others.[1] It was a traveling experience that shows how Enlightenment toleration, sociability, and living in peace with and among others also existed in Aleppo, a place many people in recent years have associated with barbarism, above all else.

Britons in the City

England's relationship with Aleppo goes back to 1583, when Queen Elizabeth I, as part of her policy of making political and economic alliances with the Ottomans, granted a few English merchants permission to trade in the Levant. The Levant Company then was established, with its factories operating in Constantinople, Alexandria, Smyrna, and Aleppo.[2] Throughout the seventeenth century, Aleppo was the main headquarters for the Levant Company, and at a time when there was great demand in Europe for Persian silk, the city, being on the overland routes between Persia and Europe, became a very attractive place for Europeans to do business.[3] European merchants continued to frequent Aleppo in the seventeenth century, but during the eighteenth century, Aleppo began to lose its reputation as a trade center. The civil wars in Persia were the main reason for the reduction of British commercial activities in Aleppo, and the Persian silk stopped appearing in the city's great bazaars and warehouse. The Levant Company in Aleppo suffered from the decline of trade in Persian silk, but it also faced another problem in the city: French competition.[4] The French used to sell their cloth in Aleppo at a much cheaper price than the English.

Despite the decline of trade in the city, many Britons kept their commercial connections with Aleppo, the third-most important city in the Ottoman Empire, after Istanbul and Cairo. Jonas Hanway, a well-known merchant and traveler in Persia and Russia, wrote about the decline of trade in Ottoman cities while he also confirmed that "Yet is not the trade entirely sunk; on the contrary it is said, that from Aleppo we annually import six hundred bales of raw silk. This alone is a very great national object" (1753, 32).[5] Although British trade with the Levant continued to decline in the second half of the eighteenth century, Britain was not prepared to quit the Levant, as Aleppo was its gateway to India: it was a commercial hub on the edge of vast deserts connecting Syria with the Persian Gulf. Aleppo was a caravan city between Iraq and Syria, and many East India Company servants and military officers visited and lived in Aleppo on their way to India.[6] These travelers knew that it was much more convenient and less expensive to pass through Aleppo than to cross the Cape of Good Hope in a period when the East India Company, during the peak of its commercial and military expansion, needed to shorten the distance between India and Britain.[7]

Throughout the seventeenth and eighteenth centuries many Britons frequented Aleppo and noted in amazement the level of diversity that existed in this trading city. Charles Robson, a fellow of Queen's College, Oxford, entered the service as a chaplain in the Levant Company in 1628. Upon his arrival in Aleppo he noted how the city was "an Epitome of the whole world. There scarce being a Nation of the old World (except that all-hated Spaniard) who hath not some trading either here or hither. English, French, Dutch, Italian, Jews, Greeks, Persians, Moores, and Indians, &c [sic]. Men of all Countries, of all Religions: Georgians, Nestorians, Cophti, Armenians, Georgians, &c [sic]" (1628, 15). In the presence of such diversity Britons, mostly single men from respectable families, were sociable creatures in the city. But the idea of excessive exposure to the difference existing in Aleppo sometimes disquieted the Levant Company. As Christine Laidlaw stated, "The company's interest lay in keeping the factory communities within the Protestant fold and safe from alien, particularly Roman Catholic influences. It did so by installing English clergymen to provide religious discipline and usual service of marriage, baptism and burial" (2010, 75).

At a time when Protestant-Catholic animosities were strong in Europe, British travelers in Syria reported that Aleppo represented the ultimate space of diversity. Further, residents of Aleppo were aware that they were bound to be entangled with the world of others, a world in which the idea of being among different peoples and cultures could lead to the change of self. One late-seventeenth- and early eighteenth-century chaplain in Aleppo, Henry Maundrell, was among the British chaplains whom the *Colonial Church Chronicle* in 1858 extolled as being the "first" who "tried to raise and confirm the religious feelings and practice of their own countrymen, and then in various ways, laid a good foundation for direct Missionary work among the Mahometan population who surrounded them" (1858, 48). This self-congratulatory statement by the Chronicle attests to the religious enthusiasm of the early English chaplains in Aleppo. Maundrell was a devout chaplain but was not a missionary in the city. He did not convert one single Muslim to Christianity; rather, he was a man of God mostly keen on keeping his countrymen within the Protestant fold. He guarded the religious "purity" of his countrymen when Aleppo was not only more diverse than England but also offered a lot of experiences for a young Englishman looking for fun, mainly sexual ventures abroad.[8]

Although he might have guarded against living a non-Christian life, Maundrell knew that Britons in the city were bound to be living and communicating with others. Arriving in Aleppo at a time when at least forty Britons lived in Aleppo, Maundrell, a churchman from Bromley, held the chaplaincy post in 1695 until his death there in 1703. He wrote to Thomas Sprat, bishop of Rochester, telling him how the English in the city were virtuous, self-disciplined, and, most importantly, committed worshippers. "I need only add, that they still continue," wrote Maundrell in his travel diary, *Journey from Aleppo to Jerusalem at Easter A.D. 1697,*

> pious, sober, benevolent, devout in the office of religion; in conversation innocently cheerful; given to no pleasures but such as are honest and manly; to no communications, but such as nicest ears need not be offended at; exhibiting in all their actions those best and truest signs of a christian spirit, a sincere and cheerful friendship among themselves, a generous charity towards others, and a profound reverence for the liturgy and constitution of the Church of England. (1703, b1)

Britons in Aleppo were charitable toward others but also loyal to their Protestant religion and friendly to their English community. One here notes that Maundrell preached the idea of staying loyal to the Protestant faith and the Church of England in a city where to be different—culturally and religiously—was the norm rather than the exception, as Aleppo, as Robson noted, included people from all over the world. Maundrell expressed the idea that to be among others and to be charitable to others does not necessarily stop you from being loyal to the Church of England and the English community. After all, it was in this cosmopolitan city, where people from all religions were free to worship, that he could pose himself as a cosmopolitan Briton, showing how the English abroad were responsible toward their national community and religion as well as toward the world of others in which they lived.

Maundrell did not recall any anecdote in his narrative indicating that his countrymen interacted with other people living in the city. But he was equally aware that Britons were living in the world of others, to whom they showed charity. English charity was not countered by noncharitable acts against the English community;[9] rather, Britons in the city were allowed to worship freely. "It is our first employment every morning to solemnize the dayly service of the church; at which I am sure to have always a devout, a regular, and full of congregation," Maundrell wrote. "In a word, I can say no more ... than this," Maundrell concluded, "that in all my experience in the world, I have never known a society of young gentlemen, whether in the city or the country ... so well disposed in all points as this" (1703, b1).

The Dutch travelers Jan Aegidius van Egmond and John Heymen, who visited Aleppo, recalled how the Muslims of Aleppo were strict in their religious commitment and worship but did not threaten Christians or ban them from following their faith: "were all the Christians and Jews to become Turks, the Grand Signior would lose one of the best branches of his revenue. Not only the Charatz [poll tax], but the Alcoran itself enjoins that everyone should be indulged in the free exercise of his religion, adding at the same time, that the Christian religion is good, and many may obtain salvation by it" (1759, 339). These two early-eighteenth-century European travelers—Egmond and Heymen—came from the Netherlands, known for its religious toleration, but were impressed by the toleration that existed in the Ottoman Empire, the world of Islam. It was in this context of cosmopolitan Aleppo that Maundrell and the rest of his English flock practiced their faith. And this happened during a period in England when men of God like the dean of Norwich, Humphrey Prideaux, propagated ideas about the violence of Muslims toward Christians.[10]

Maundrell's commitment to his religion and community abroad occurred in a city that tolerated and lived with differences. The Scotsman Alexander Drummond was in the Levant in 1745, a year when his country, Scotland, was going through the Jacobite Uprising, an uprising against King George II of Great Britain that threatened the union between England and Scotland. It was also the year when England was at with war with France over the Austrian succession. In this difficult year for Britain Drummond defined his Britishness: one of loyalty to the British monarch in the context of being among others in Aleppo—Muslims and other Europeans living in the city. In the several letters that appeared in his book, *Travels through Different Cities of Germany, Italy, Greece and Several Parts of Asia as Far as the Banks of the Euphrates*, Drummond recalled meeting the Muslim governor of the city. "I explained the causes of the war" in Europe, Drummond wrote,

> inveighed against the unlimited ambition and insincerity of our enemies; expatiated upon the kingly and paternal virtues of our monarch, whom I stiled his Imperial Majesty of Great Britain; described his naval power, and concluded with saying, that were it not for the singular regard he had for his faithful ally, the most sublime sovereign of the Ottoman Empire, in a few months, not a French ship would dare to shew her colours within the seas of Egypt, Asia, or the Archipelago. (1754, 186)

The enemies of Britain here were France and her allies, not the Ottoman Sultan.[11]

In Aleppo Drummond expressed a nationalist sentiment. But he was also sociable and friendly with the people he encountered in the city: the Muslim governor as well as the French and Dutch consuls. "At present," he wrote, "Arthur Pollard, esquire, very worthily fills the chair of consul; and to him I have very great obligations, as well as to Mr. Hiermans, consul for the United Provinces, who lives in perfect harmony with the Englifh gentlemen: I likewise owe great civilities to Monsieur Delane, the French consul, in consequence of a letter he received in' my favour, from brother Lemaire of Cyprus" (1754, 185).

There Britons appeared steeped in cosmopolitan practices, although they sometimes expressed narrow nationalist and religious confessions. When, in 1760, the Armenian nationalist Joseph Emin, a friend of Lord Bolingbroke and Edmund Burke in England, tried to recruit the Armenians and Turcomen tribes to take possession of Aleppo, two Britons—Patrick Russell, a Scottish doctor, and David Hays, a merchant, then residents in Aleppo—dissuaded him from undertaking such a dangerous task. For these two Britons, Emin was a good Christian, but his nationalist aspiration of securing Armenian independence from the Ottomans was not necessarily something they agreed with. In a letter published in the first volume of Emin's memoir, edited and published by Emin's granddaughter Amy Apcar, Russell wrote to the earl of Northumberland, explaining to him why he and Hays prevented Emin from doing anything that would disturb the religious and cultural harmony in Aleppo: "The earl of Northumberland has great merit in finding out Emin by his lordship's surprising him, and in patronaging him who is really worthy of esteem from every man of spirit. If he had not hearkened to us, the consequences of his enterprise would unavoidably have been fatal to all the Christian subjects of the Othman empire; nor could Europeans have been prevented from sharing their fate" (1918 [1792], 159).

Emin's design to disturb the harmony existing in Aleppo did not impress Russell and Hays. Russell was, as Emin noted, a man so respected in Aleppo that he was allowed, unlike other Christians and Europeans in the city, to wear a white turban, a privilege only men from the elite enjoyed. Similarly Hays had considerable knowledge of the local culture in the city. His cosmopolitanism appeared in the way he raised his daughter Marianne Hays, later the wife of the famous British consul in Aleppo John Barker. As John Griffiths, who visited Aleppo in 1785, noted, Marianne spoke five languages: "at the tender age of seven she spoke fluently Arabic, Greek, French, Italian and English languages" (1805, 335).

As this indicates, Britons in Aleppo in the second half of the century saw themselves as members of the social and religious fabric of the city and were not keen on disturbing the religious harmony that existed there. When meeting the nationalist Armenian Christian in Baghdad in 1744, Abraham Parsons, the consul of the Levant Company in Iskenderun, was not impressed by his character, calling Emin "an imposture, who had passed himself off in London some few years back as Prince Heraclius, of Georgia" (1808, 139). Emin also recalled in his diary how Parsons was not sociable to him, a fellow Christian in the East. Parsons, unlike Emin, favored diversity and mixing rather than the religious and national homogeneity Emin advocated.

Unlike all previous residents in Syria, Parsons had a special affinity with Aleppo. As a consul in Iskenderun, which is a one-day journey from Aleppo,

Parsons often visited the commercial "capital of Syria." "This is the fourth tour which I have made from Scanderoon to Aleppo," Parsons declared, "and having little else to do but to walk about and make observations, it has enabled me to see more than other persons who have resided here many years, as they had business of more consequence to take off their attention" (1808, 56). Parsons observed the bazaars, guilds, and castle of Aleppo, recalling that "This is by all (who have seen the capital of other great cities in Turkey) allowed to be the best built, handsomest, and cleanest capital of any city in the grand signior's dominions, and next to Constantinople for size and number of inhabitants being the capital of Syria" (1808, 56). Further, Parsons visited the suburb of Judaida in Aleppo and noted that "all the Jews and Christian of every dominations" lived there: "some of them are very rich, and have superb houses, with large gardens, hot and cold baths, and every thing corresponding, either for convenience and luxury" (1808, 56–57). Christians and Jews mainly lived outside the walls of the old city of Aleppo, but the rich among them owned shops and warehouses inside the city. Some of them were the richest among the city's merchants.

Parsons was impressed with Aleppo's diversity. In Aleppo, he wrote, "there are four denominations of native Christians, who are called the four nations, viz. the Roman Catholics, Greeks, Armenians and Syrians" and that "Here the Europeans, or Frank merchants, are under the protection of the consuls of the nation to which they belong, which are four in number, French, English, Venetian and Dutch, all of whom, with their servants, are permitted to reside in the city" (1808, 57). Parsons also found many Catholic missionaries living in Aleppo: "Here are four convents of missionaries, under the protection of the French consul; one convent is denominated La Casa di Terra Santa, or the House of the Hold Land; the second is the order of Saint Francis: the third are Carmelites; and the fourth Jesuits; in which last there remains only one Jesuit, who notwithstanding the Pope has abolished the order, declares he neither quit the convent, nor the dress of the order during life" (1808, 57–58). The Jesuits were banned in Enlightenment Europe, but in Aleppo, Parsons found, they were allowed to assume their habits and worship freely. What fascinated Parsons in this diverse cultural and religious landscape was that the Pasha of the city mainly cared to collect taxes from the non-Muslims, while leaving it to community leaders to manage the affairs of their communities. Parsons never recalled any form of religious persecution there.

Non-Muslims were not only allowed to worship freely but could also participate in the economic life of the city. In a city where "there are to be seen immense quantities of the richest goods from India, Constantinople, Smyrna, Damascus, and other places, beside the various manufactures of Aleppo," Parsons found, "The weavers" could be "Turks who live in the city, or Jews and Christians who live in the Jewdeda [suburb]" (1808, 60–62). The newly appointed ruler, Muhsinzade Mehmed Pasha, was previously a grand vizier in the Ottoman court and sent by Sultan Mustapha III to curb the rebellion of local factions and restore political order in the city. Pasha supported trade with Europeans and was friendly to the European consuls living in the city, mainly the French and the English. "I am told that among Christians there are many different sects," Parsons quoted the Pasha as saying,

> and that each has a different way of worshipping God, and that the French and the English differ much: I do not pretend to know who is most right, but must observe, that if eighteenth French men must have upwards of thirty of the religious men of your church to superintend their conduct, and that twelve English men can be kept in order by one religious man's of their's, I must certainly give the preference to the English church; and "if I will be Christian" (added he, smiling) "I will be of their church" (59).

The reference here is to the chaplains and missionaries accompanying the European merchants and consuls in Aleppo, mainly the French. The presence of large numbers of these men in Aleppo attest to the Ottoman governor's policy of tolerance, allowing Christian Europeans to worship freely in his land.

Thus, Europeans were free to worship in Aleppo as well as, Parsons observed, to wear European clothes and participate in the entertainments and pleasures the city's elites often practiced—that is, hunting. "In autumn, winter, and spring, there is plenty of game within the distance of four or five miles from the city, where the franks go without interruption from the Turks" (1808, 66). Living in Aleppo, a wealthy and prosperous city, Parsons concluded, was itself an encouragement for Europeans to foster peace and harmony among members of their own community. "Upon the whole, the French, English, Italians, and Dutch, live as comfortably ... as there is always a good harmony subsisting between them, and even if their countries are at war at home, they not only live peaceably, but amicably there" (66).

Thus, British travelers in Aleppo found a harmonious space where Europeans lived comfortably without feeling threatened by Muslim persecution. As Richard Pococke, an eighteenth-century English traveler in the Levant, wrote, "The English pass their time here very agreeably; and in the excursions which they make for pleasure they are commonly respected by the Arabs, Curdeens and Turcomen, there being very few instances of their having been plundered by them. They live very sociably with one another, and pass two or three days in the week either in the gardens, or under a tent in the country, or else amuse themselves in the season with country diversion" (1737, 152).

Aleppo was a cosmopolitan place for another Briton who arrived there as a boy at the age of ten in 1774, the same year Parsons visited the city. Henry

Mohammad Sakhnini

Abbott was born in Pera, a suburb of Constantinople, to a family of established Levant merchants. After his father died, the family decided he should go to Aleppo to learn from his uncle Robert Abbott, an English merchant based in the city, the skills needed to set up as an independent merchant. "[At] the age of 15," wrote Henry in his unpublished memoir about his upbringing in the Middle East and his life in Bengal, "my uncle took me into his counting house that I might learn merchants accounts & the general routine of commerce" (n.d. 9). The Abbots were an established family in Aleppo: John, Henry's second uncle, was the British consul in the city and was greatly respected by the local Arabs and by the English travelers in the region. Besides receiving professional training as a merchant, Henry was delivered to the care of the company chaplain, Mr. Robert Foster, who served as a new instructor for the young Henry. Forster, "a man of education & talent," Henry wrote, was keen on "discharging trust reposed to him to the best of his abilities, both on my literary studies & moral principles" (n.d. 9).

Abbott never mentioned whether he was asked to go to church; rather, Forster was mostly concerned about Abbott's literary studies and morality. Abbott's secular education in Aleppo signaled a move from the strict religious commitment that Maundrell recalled earlier in the century. Aleppo here was a cosmopolitan space where a young Englishman learned Arabic and Italian, speaking with the locals in Arabic on his frequent riding trips to the deserts surrounding Aleppo.[12] He also spoke in Italian, the lingua franca in Aleppo throughout the period, with Dr. Salina, "a very respectable man who had two daughters & three sons grown up all of whom were fond of me, & as I generally met good company there, I passed my evenings very agreeably with them when I was not otherwise engaged" (n.d. 11). Abbott spent many evenings among the European community in Aleppo, organizing group readings and play acting. "The language we spoke was always Italian," Henry added,

> the young ladies with great good will corrected my errors. We often read together & frequently got up plays from Mitastasio & Goldoni in which I always had a part, & whenever we were sufficiently perfect, my uncle the Consul generally invited a party of select friends to form the audience which we exhibited upon a neat little theatre he had built for the purpose at his own house. The characters in general well supported, & the whole together formed at once, a pleasing, innocent & instructive amusement, much to be desired amongst a circle of friends in any part of the world. (n.d. 12)

Abbott's account about the scenes of diversion that Europeans sought in the city was confirmed by James Capper, who visited Aleppo in the late 1770s, who wrote, "The Europeans by a general subscription have built a small theatre, which they have fitted up with great taste" (1783, 178). "During the winter season they perform French and Italian comedies, and even sometimes attempt operas with very great success," he added (1783, 178).

Eighteenth-century British travelers and residents in the city were free to trade, socialize among themselves, and have their own dinners, parties, and drinks without fearing that Muslims would interfere in their private affairs. These innocent and amicable relationships among the Europeans in Aleppo, among which the English played a major part, were emphasized by Alexander Russell. A long-term resident in Aleppo between 1740 and 1754, Russell was a Scottish physician whose detailed observations of Aleppo, initially published in 1754, were read by almost every English and British travel writer who wrote about the city thereafter.13 Russell's book, The Natural History of Aleppo, first published in 1754 and later revised by his brother Patrick in 1794, is one of the major sources on the Ottoman city of Aleppo.

Among the many topics covered, Russell noted how Aleppo accommodated many Europeans, both those who resided in the city as well as just visitors. The "Franks," as the locals called them, included "English, French, Venetian, Dutch, and Tuscan, or Imperial, subjects" (1794, 1). Russell found that Europeans did not feel pressured to learn Arabic. "Of the Europeans, even those who live long in the country, very few acquire more knowledge of Arabic, than is barely sufficient for familiar conversation, and it is very rarely than any of them take the trouble of either learning to read or write it" (1794, 2). Nor did the "Franks" feel pressured to conform to the local dress code: "The Consuls and several of the private gentlemen, retain the European dress" (1794, 2). "It was formerly the custom of all, or most of the Franks," Russell recalled, "to wear the Turkish dress, retaining the hat by way of distinction; but of late, the far greater part of the English dressed in the European fashion" (1794, 2). Even the Catholic missionaries in Aleppo, mostly French Catholics, Russell marveled, wore "the proper habit of their order" (1794, 7). In a period when few foreigners turned up on the European shores, Aleppo, as Russell noted, tolerated foreign habits.[14]

Europeans felt free to "entertain reciprocally; they have card parties, weekly concerts, and sometimes, in the Carnival, masquerades," Russell added (1794, 12). "Neither competition in trade nor the intervention of national ruptures in Europe, broke off the social intercourse in Syria" (1794, 12–13). The English in Aleppo, Russell noted, socialized with other Europeans. "The tables of Europeans," Russell added,

> are well supplied with provisions of all kind, except sea fish, which can only be procured fresh in the winter. The cooks, as well as most of the

other menial servants, are Armenians, but have been taught French or English cookery, and only now and then, by way of variety, serve up some of the country dishes. ... The wines in common use are the dry white wine of the country, and a light red Provencal wine. The French present Liqueurs at the Desert. The English drink a draught of very weak punch, before dinner and supper; a custom found so deliciously refreshing, that most of the other Europeans, many of the native Christians, and some even of the Turks have adopted it. (10)

In Aleppo Britons often found themselves immersed in the world of others, interacting and socializing with them rather feeling superior to them.

For Russell, Britons and other Europeans in Aleppo were so close to each other that they resembled the members of one family. "The Franks, in general live together in harmony" (1794, 12). At a time of war between England and France, "public ceremonies between the Consul were suspended. But the private relation of men brought together by accident in a distant country, whom choice had led to form friendly connections, still remained sacred" (1794, 13). Political problems in Europe remained in Europe. "Individuals continued to visit and amuse themselves as usual; politics were banished from conversation, by mutual consent," Russell remarked, "and without forgetting that they owed to the public cause, both parties, while they wished for peace, continued to remember what in the meantime might be conceded to civility, and private friendship" (1794, 13). Europeans cultivated peace among themselves, and in so doing they offered to their Muslim neighbors a certain interpretation of their European-ness, one highlighting peace and amity. "A Missionary, describing the ceremonial visit made by Europeans at the annual [Muslim] feasts," Russell reported,

> justly remarks to his correspondent, that he need not be surprised at those mutual civilities among people of different countries, for the French, English, Italian, and Dutch, in respect to the people among whom he dwelt, considered themselves as persons of the same country, and in that light, were viewed by the natives, who, without distinction reckoned them all Franks (1794, 14).

Russell, a Scot, appears to have been comfortable seeing all Europeans interacting as one people, belonging to the same country. Aleppo appears as the medium through which Russell developed a sense of European identity, one of peace and harmony with fellow Europeans.

But this sense of European unity in Aleppo did not necessarily exclude non-Europeans but also included Muslims. When Europeans gather outside the city for "hunting and shooting parties," Russell wrote, Europeans "are sometimes visited by the Emeer, or king of the Arabs, in his way to, from the city" (1794, 16–17). Immediately upon welcoming him in their tent, the Europeans showed "great civility, and together with his retinue, (which seldom exceed five or six persons) is treated with wine, or spirits, either being more agreeable to the Emeer than coffee" (1794, 17).

Russell noted the sense of amity that existed between different groups in Aleppo, stating, "[T]he several Frank nations at Aleppo are equally protected by the government; and the privileges they enjoy are very considerable. The consular houses are respected as sanctuaries; the officers of justice cannot enter even the houses of private Merchants, without permission," remarked Russell (1794, 19). Moreover, the wealth they made in this tolerant city allowed them to bring from Europe material and objects that reminded them of their nations. For example, they brought furniture from England into their Aleppo houses. Julius Griffiths, an English doctor, visited the house of his friend Mr. David Hays, an Aleppo merchant, in 1785. He found that "On the right hand of the square the first apartment was a dinning-parlour, generally furnished in the English style" (1805, 335). But this English gentleman, Griffiths noted, was not strictly "Englished" in the way he designed his house; Hays was a cosmopolitan man who incorporated into the design of his house some local Alepine style. Like the houses of the upper classes in Aleppo, Hays included a "fountain constantly playing" in the yard and in the divan "a marble fountain in the middle, and cushions upon a raised platform along three of the sides" (1805, 335). Griffiths was fascinated with what he called this "luxurious retreat" that was "fitted up with much taste" (1804, 335). European residents also often used articles of luxury from the East. "For most of their houses," wrote Francis Vernon, another British traveler in Aleppo toward the end of the century, "were elegantly furnished, with carpets, sophas, cushions, &c, the manufacture of Persia" (1792 161). The material objects that their countrymen adopted from the locals fascinated Vernon and Griffiths. Observing the plants on display in Hays's house, Griffiths noted how "The left side was a wall securing the premise from the street; against which a latticed woodwork supported the most fragrant jasmines, roses, and other flowering shrubs, which were continued in front of the servants' offices, near the door of entrance; rendering the whole a delightful place of residence" (1805, 335).

Griffiths was intrigued by all the possibility of luxury that Aleppo offered the English. "The style of life differs from that adopted in England," he wrote, "and a kind of lassitude prevails, to which the hardier inhabitants of the north is a total stranger" (1805, 539). The routine of life in commercial Aleppo was less stressful than it was in commercial England: "An early breakfast of coffee, served without ceremony, precedes the pipe; in attention to business . . . or a short walk, or in absolute inactivity, pass the hours until the hour of dinner, which is in general more frugal than supper" (1805, 539). "After dinner, the siesta, or afternoon's nap," Griffiths added,

> is resorted to, and shorten the time until the ladies and gentlemen assemble for the evening, by invitation or otherwise; when a well supplied table, and social conviviality, detain the guests till nearly midnight, when each returns home. (539)

For Griffiths, English expatriates were not only working and making profits in Aleppo but also maximizing the opportunities of pleasure, "conviviality," and conversation the city offered to its rich inhabitants.

In *The Modern Syrians: Or, Native Society in Damascus, Aleppo, and Other Mountains of the Druses* (1844), Andrew Archibald Paton, who lived in Damascus and Aleppo between 1841 and 1844, recalled how his countrymen who visited and lived in the city during the eighteenth century were men of sociability. Paton examined the archives of the British factory in Aleppo and found an account—he did not mention who wrote it—about the English captain Francis Vernon, who visited Aleppo in the late eighteenth century. Once "in carnival time," the French were having "a pic-nic party" where they were all dressed up. Vernon, a friend of the Aleppo elites, borrowed the governor's clothes and stormed the party without informing the French consul that he planned to visit him. On the spot, a messenger was dispatched to inform the French consul about the sudden visit, and the French ambassador felt embarrassed. At previous occasions he would often wear formal European clothes when meeting a Muslim man of authority, but now he was in the middle of a masquerade and, thus, was not prepared to meet the Muslim governor. "Harlequin," as Paton described the French consul,

> wishing the Pasha at the bottom of the koik [river in Aleppo], stumbled out an expression of the delight which his visit would give him. The nation, his subjects, aggravated his torments by their suppressed laughter; so he cast away mask and sword, and borrowing a wig from one, a cocked hat from another, and covering harlequin's jacket with a cloak, begged them to suspend their merriment until the redoubtable visit was terminated. But in vain did he arrange his wig, and frown them into decorum; the cloak, however skillfully spread out, still left exposed a pair of legs resplendent with all the colours of the rainbow, and an uncontrollable burst of laughter rung round the tent; just at this moment the dragoman re-entered: —His Excellency is come. The Consul is at his last gasp, when Mr. Vernon and suite make their appearance, and join their laughter to that of the French. (1844, 256)

In Aleppo Vernon impersonated a Muslim man of authority to poke fun at the French living in the city. But this anecdote also revealed that Vernon, who, like Henry Abbott, came from a family with a long history of trade in the Ottoman Levant, was friendly with the Muslim governor as well as the French living in the city.

Patton's anecdote about Vernon also shows how Britons and other European residents in Aleppo felt free to hold carnivals and masquerades. Britons participated in these acts in the context of friendship and amity with the local Muslims, the governors, and the inhabitants. In his travel book, *Voyages and Travels of a Sea Officer,* Vernon recalled his fascination with the European Christmas carnival in the city, mentioning how Europeans passed the time there with "one continued circulation of entertainment, at the French, Venetian, Dutch, or English consuls; where the society of many respectable European families was heightened by the splendour of Eastern magnificence" (1792, 161). Europeans in Aleppo were free to have their own parties where men and women mix, a practice that did not exist among the Alepine society, including the local Jews and Christians. "We frequently formed balls and masquerades, thereby giving an opportunity for the ladies to convince us, they were by no means deficient in polite accomplishments" (1792, 161).

Europeans were free to mix among themselves and pursue their European habits with no Muslim intervention in their affairs. Donald Campbell visited Aleppo in the same year Vernon was there; he was a Scot of noble descent and a lieutenant in the British army in Madras. He arrived in Aleppo in 1785 and wrote his observations of the city in travel book titled *A Journey Overland to India Partly by a Route Never Gone Before by Any European* (1795). Campbell noted that a French merchant in the city hosted him: "A gentleman of the opulence and consequence ... with a house such as I have described, and a disposition to social enjoyment, was not, you will conclude, without a resort of company and friends" (1795, 55). "Parties of pleasure had no intermission," Campbell added, "while I was there; —and as the ladies of Europe or of European extraction in that country are highly accomplished, speak many languages, are indefatigable in their efforts to please, and receive strangers from Europe with a joy and satisfaction not to be described" (1795, 55).

Campbell did not restrict himself to the European parties in the city; he was also keen on learning about the local culture. He accompanied his host, the Frenchman, to the coffeehouses of Aleppo, where they listened to the Hakawati, the storyteller of the city, and attended one famous shadow theater in Aleppo known as Karakouz and Hacivat. It was in Aleppo that Campbell learned, after watching these shows and conversing with his French friend who knew Arabic, that "KARA-GHUSE had from time to time created a great deal of uneasiness, not only to private offending individuals, but also to the magistracy itself, that no offender, however entrenched behind power, or enshrined in rank, could escape him—that Bashaws, Cadi's, nay the Janissaries themselves, were often made the sport of his fury" (1795, 73). Even the ordinary people in Aleppo respected him "(as we venerate the liberty of the press) as a bolder teller of truth, who with little mischief does a great deal of good, and often rouses the lethargic public mind to a sense of public dangers and injuries" (73). At a time when the authorities in both Pitt's England and republican France were narrowing the space of a free press, Campbell recalled how Aleppo allowed dissenting voices in its public sphere. "[I]f Master Karaghuse," the Frenchman recalled, "was to take such liberties in France, Spain, Portugal, or Germany, all his wit and honesty would not save him from punishment" (1795, 73).

Conclusion

In this article I have shown that eighteenth-century British writings about Aleppo were as diverse and complex as the city itself. In the entanglement between Britons and the differences that existed around them, Aleppo acted as a testing ground for some eighteenth-century British cosmopolitan experiences. Britons did not necessarily pose themselves as superior to Muslims or emphasize the persecuting nature of the form of Islam practiced in Aleppo. But the level of diversity and tolerance the British experienced in Aleppo during the eighteenth century may tell us something about the present. The jump from the eighteenth century to the twenty-first century might be anachronistic and wildly ambitious, but the few refugees from Aleppo I met in London in the previous months through my modest involvement in volunteering work expressed concern about the rise of Islamophobia in post-Brexit London. Here I explored how Aleppo's cosmopolitan past and murderous present should give us a remainder that cities—however cosmopolitan and diverse they can be—are as fragile as a spider's web.

Acknowledgments

I would like to thank Professor Gerald MacLean for reading and commenting on an early draft of this article. Also, I would like to thank students and colleagues in the Globalisation: History, Politics and Culture programme at the University of Brighton whose conversations and work inspired me to research and write this article.

In The Eyes of Some Britons

Mohammad Sakhnini pursued his graduate studies at the University of Exeter, UK. He taught at the universities of Exeter, Sussex and Brighton and held a research fellowship at the University of Linnaeus, Sweden. Currently he is a visiting fellow at the Woolf Institute, Cambridge.

Notes

1. Bruce Masters (2001), Philips Mansel (2016), and Abraham Marcus (1992) demonstrated that Aleppo, during the seventeenth and eighteenth centuries, was known for its religious diversity and coexistence rather than sectarian violence, at least in the period predating the rise of a European-style nationalism in the region. In the context of economic alliances and self-interest, as Masters (2001) showed, many Muslim, Christian, and Jewish residents in the city crossed religious barriers. Also, Mansel recalled that "long periods of coexistence" between Muslims, Christians, and Jews, "although less likely to provide events to record, were most characteristic of the city than moments of violence" (2016, 27).
2. On the formation and operation of the Levant Company, see Davies (1976), Laidlaw (2010), and Wood (1953).
3. The English prelate Richard Pococke visited Aleppo in 1738 and noted, "Aleppo is the great mart of all Persian goods, especially raw silk; a large caravan comes from Balsora or Bosra, on the Euphrates, which is usually a month on the road. This trade has however much decayed since the Persian war" (1754, II:151).
4. A good account of the period explaining why the English trade declined in the Ottoman Empire can be found in Porter (1771, 396–406).
5. Masters remarked, "Trade with Europe diminished, but obviously did not come to an end with the withdrawal of European factors from the city. There was still a demand for European manufactured goods in Syria and corresponding European demand for northern Syrian's raw cotton and silk" (1991, 49).
6. On Aleppo as a caravan city, see Mather (2011, ch. 1).
7. See Carruthers (1928, introduction).
8. Alexander Russell, who lived in Aleppo between 1740 and 1752, noted how in one quarter of the city "prostitutes are licensed" by the governor "whom they pay for his protection" (1794, II:263).
9. Maundrell's emphasis on the idea that Protestants in Aleppo were men of religious piety indicates the tolerance he found in Aleppo, but it also clearly engages his text with a significant concern that emerged during the Enlightenment: the tradition of toleration that appeared in John Locke's *Letter Concerning Toleration* (1689), a text that argued for the separation of personal faith from political government while also finding that one group in England, the Roman Catholics, cannot be tolerated because of their allegiance to foreign Catholic rulers. Maundrell in Aleppo, in a subtle way, was a product of the Enlightenment tradition of toleration but—as his experience in Aleppo showed—the Enlightenment was also happening outside Europe.
10. Writing of what he considered the heart-wrenching divisions among Christians in the East so that his coreligionists in the West would learn a lesson, Prideaux explained how the "Saracens," as he called the Muslims, rose in the past—and still in the present—against Christians. "And having fixed that Tyranny over them [Christians]," wrote Prideaux, "which hath ever since afflicted those Parts of the World, turned everywhere their Churches into Mosques, and their Worship into an horrid Superstition; and instead of that Holy Religion which they had thus abused, forced on them that abominable Im-

Mohammad Sakhnini

posture Mahometism, which dictating War, Bloodshed and Violence in matters of Religion, as one of its chiefest Virtues" (1697, viii–ix).
11. Several scholars, notably Colley (1993), wrote profusely about British hostility to the French throughout the eighteenth century, suggesting that such an attitude reflected the rise British nationalist sentiments. This is certainly true as some British travelers in the Levant during the period openly expressed their antagonism to the French. But little in the scholarship about the meanings of Britishness during the period recalled how Britons in cosmopolitan Ottoman cities such as Aleppo also developed a sense of Britishness that was anti-French but also friendly to Ottoman Muslims, particularly the men of higher offices.
12. `Besides the unpublished manuscript about his life in Aleppo and India, Abbott published a travel account about his journey from Aleppo to India. He wrote passionately about the Arabs he met across the deserts, describing them as "hospitable, even to excess"(1789, 13). Abbott was a true cosmopolitan Briton in the East.
13. On the Russell brothers in Aleppo, see Boogert (2010).
14. On how Muslims tolerated Europeans' style of dressing up in Aleppo, Andrew Archibald Paton, a traveler in Syria between 1840 and 1843, wrote, " that the Moslems of Aleppo, unlike those of Damascus, have been so long accustomed to Franks, that the European dress would not impede my intercourse with them" (1844, 225).

References

Abbott, Henry. 1789. *A Journey on a Trip from Aleppo to Bussora*. Calcutta: From the Press of Joseph Cooper.
Abbott, Henry. n.d *Memoirs and Diary of Henry Abbott*. The British Library, India Office Select Materials MSS EUR B412 (A).
Boogert, Maurits van den. 2010. *Aleppo Observed: Ottoman Syria through the Eyes of Two Scottish Doctors, Alexander and Patrick Russell*. Oxford: Oxford University Press.
Campbell, Donald. 1795. *A Journey over Land to India Partly by a Route Never Gone Before by Any European*. London: Printed for Cullen and Company, No. 54 Pall-Mall.
Capper, James. 1783. *Observations on the Passage to India through Egypt, and across the Great Desert with Occasional Remarks on the Adjacent Countries, and also Sketches of the Different Routes*. London.
Carruthers, Douglass. 1928. "Introduction." *The Desert Route to India: Being the Journal of Four Travellers by the Great Desert Caravan Route between Aleppo and Basra 1745–1751*. Edited by Douglass Carruthers. London: Hakluyt Society.
Colley, Linda. 1993. *Britons: Forging the Nation 1707–1837*. New Haven, CT: Yale University Press.
Davies, Ralph. 1976. *Aleppo and Devonshire Square: English Traders in the Levant in the Eighteenth Century*. London: MacMillan.
Drummond, Alexander. 1754. *Travels through Different Cities of Germany, Italy, Greece, and Several parts of Asia, as Far as the Banks of the Euphrates*. London: Printed for W. Strahan.
Egmond, Jan Aegidius van, and John Heymen. 1759. *Travels through Parts of Europe and Asia Minor*. London: Printed for L. Davis and C. Reymers.
Emin, Joseph. (1792) 1918. *The Life and Adventures of Joseph Emin 1726-1809 Written by Himself*, Edited by His Great-Great Granddaughter Amy Apcar. Calcutta: Printed and Published by the Baptist Mission Press, Lower Circular Road.
Griffiths, John. 1805. *Travels in Europe, Asia Minor, and Arabia*. London.
Hanway, Jonas. 1753. *An Answer to the Appendix of a Pamphlet, entitled Reflections Upon Naturalization, Corporations and Companies, &c. Relating to the Levant Trade and*

Turkey Company, As this Subject is Occasionally mentioned in Hanway' Travels. London: Printed for R. Dodsley.

Laidlaw, Christine. 2010. *The British in the Levant: Trade and Perceptions of the Ottoman Empire in the Eighteenth Century*. London: I. B Tauris.

Locke, John. 1689. *A Letter Concerning Toleration: Humbly Submitted, &c*. London: Printed for Awnsham Churchill, at the Black Swan at Amen-Corner.

Mansel, Philip. 2016. *Aleppo: The Rise and Fall of Syria's Great Merchant City*. London: I. B. Tauris.

Marcus, Abraham. 1992. *The Middle East on the Eve of Modernity: Aleppo in the 18th Century*. New York: Columbia University Press.

Masters, Bruce. 1991. "Aleppo: The Ottoman Empire's Caravan City". In *The Ottoman City Between East and West: Aleppo, Izmir and Istanbul*, edited by Edhem Eldem, Daniel Goffman, and Bruce Masters, 35–43. Cambridge: Cambridge University Press.

Masters, Bruce. 2001. *Christian and Jews in the Ottoman Arab World*. Cambridge: Cambridge University Press.

Mather, James. 2011. *Pashas: Traders and Travellers in the Islamic World*. New Haven, CT: Yale University Press.

Maundrell, Henry. 1703. *A Journey from Aleppo to Jerusalem at Easter, A.D. 1697*. Oxford: Theatre.

Parsons, Abraham. 1808. *Travels in Asia and Africa Including a Journey from Scanderoon to Aleppo, and over the Desert to Bagdad and Bussora*. Edited by John Paine Berjew. London.

Paton, Andrew Archibald. 1844. *The Modern Syrians; Or, the Native Society in Damascus, Aleppo, and the Mountains of the Druses from Notes Made in Those Parts during the Years 1841–2–3, by an Oriental Student*. London: Longman, Brown, Green, And Longmans, Paternoster-Row

Pococke, Richard. 1754. *A Description of the East and Some Other Countries*. Vol. II. London: W. Bowyer.

Porter, James. 1771. *Observations on the Religion, Laws, Government and Manners of the Turks, the Second Edition, Corrected and Enlarged by the Author to Which Is Added the State of The Turkey Trade from Its Origin to the Present Time*. London: Printed for J. Nourse.

Prideaux. Humphrey. 1697. *The Nature of Imposture Fully Displayed in the Life of Mahomet with a Discourse Annexed Vindicating of Christianity from this Charge; Offered to the Consideration of the Deists of the Present Age*. London: Printed for William Roger.

Robson. Charles. 1628. *News from Aleppo*. London.

Russell, Alexander. 1794. *The Natural History of Aleppo: A Description of the City, and the Principal Natural Productions in Its Neighbourhood, Together with an Account of the Climate, Inhabitants, and Diseases; Particularly the Plague*. London: Printed for G. G and J. Robinson.

The Colonial Church Chronicle And Missionary Journal. 1858. London: Rivingtons, Waterloo Place.

Vernon, Francis. 1792. *Voyages and Travels of a Sea Officer*. Dublin: Printed for Wm. M'kenzie.

Wood, Alfred. 1953. C. A. *History of the Levant Company*. Oxford: Oxford University Press.

Chapter 10

An Ordinary Place
Aboriginality and 'Ordinary' Australia in Travel Writing of the 1990s

Robert Clarke

While much travel literature exploits readers' desires for the exotic and strange, it would seem that 'ordinary' people, places and experiences have become increasingly popular subjects in Anglophone travel writing. Whether it be a domestic setting in the south of France or Tuscany, or a tour of the pubs of Ireland, or a search for "tranquillity, stability, and conventionally civilized values" (Lawson 1993: 8), the everyday and even banal have become frequent sources of inspiration for many contemporary travel writers. The turn to the ordinary may relate to the sense of exhaustion or belatedness that Patrick Holland and Graham Huggan diagnose as a central rhetorical feature of contemporary travellers' tales (Holland and Huggan 1988). It might also be evidence of the closer entwinement of travel literature with the tourism industry: as exotic locales become harder to find, the everyday becomes a potential source of 'novelty' to sate the touristic gaze. And it may reflect the democratising trends observed in other fields of postmodern cultural production, such as radio and television (see Bonner 2003; Scannell 1996).

On the other hand, the turn to the ordinary and quotidian in travel writing may reflect the cultural values of a given place. Australia, for example, promotes itself as a society and culture that celebrates the ordinary and everyday. The Australian tourist industry, to be sure, trades on the exotic appeal that the continent offers international visitors. However, for travel writers in the 1990s, like Bill Bryson, Mark McCrum and Annie Caulfield, ordinary Australia was just as enticing as its exotic counterpart. Yet the ordinary like the exotic is a construct open to competing ideological claims. The exploration of ordinariness may signal a demotic turn in contemporary travel culture. But in postcolonial cultures such as Australia, the experience of the ordinary may also bring the traveller into contact with the disturbing legacies of colonialism. This is especially notable when the traveller seeks out Australian Aboriginal culture in the context of everyday life.

Notes for this chapter begin on page 185.

Recent Australian travel narratives by foreign travellers are distinguished by the way they represent encounters with Indigenous Australians and their cultures.[1] While there is ample evidence of the gross appropriation of Aboriginal culture (see Clarke 2009), new aesthetic regimes have developed within the genre that ostensibly eschew the tropes of the exotic to focus on encounters with Aboriginality within the context of 'ordinary' Australia. 'Aboriginality' refers not just to Aboriginal people and their culture, but more generally to a field of discourses grounding everyday interactions and dialogues among and between Aboriginal and non-Aboriginal Australians (see Langton 1993). It refers particularly to the way Aboriginal people are engaged with and talked about by non-Aboriginal Australians; and to how national and local histories of race relations are either admitted to or excluded from public consciousness, and the effects of this.

The 1990s was a particularly turbulent period in the history of Australian race relations. It was a decade that witnessed, after lengthy legal challenges, recognition first by the High Court and then by the Commonwealth government of the continuing rights of Aboriginal people to their lands, their 'native title'; the implementation of a formal, government-sponsored program for reconciliation between the Indigenous and non-Indigenous communities; the revelations and recommendations of the *Bringing them Home* Report (Human Rights and Equal Opportunity Commission 1997) that provided personal and collective accounts of the State programs that oversaw the forced removal of an estimated 1 in 10 Aboriginal children (the 'Stolen Generations') from their families over most of the twentieth century;[2] and Prime Minister John Howard's refusal to issue an apology on behalf of the nation to Aboriginal people for such practices.[3] It was also a decade during which Aboriginal art and culture achieved widespread international prestige, and a period in which the institutions of the Australian travel and tourism industries co-opted Aboriginal culture in their campaigns to attract ever more visitors to Australia.

This paper is concerned with a number of travel books that seek ordinary Australia and find, through encounters with Aboriginality, a place and culture far removed from either the stereotypes of the tourist brochures or the friendly and quirky characters that inhabit the soap operas and films that have been profitable exports of Australian popular culture. For instance, books like English writer Mark McCrum's *No Worries* (1997), *The Winners' Enclosure* (1999) by Anglo-Irish author Annie Caulfield, and *Down Under* (2000) by U.S. travel writer Bill Bryson, explicitly critique the treatment and representation of Aboriginality in everyday Australian life and disturb two important myths of contemporary Australian identity and multiculturalism, promoted through tourism and popular culture.[4] The first is that 'ordinary' Australia is a place where, in Meaghan Morris' words, "difference isn't *inherently* threatening"

(Morris 1993: 254). The second is that Australian space is a privileged site for the nostalgic performance of *white* – specifically Anglo-Celtic – ethnicities; of re-capturing a sense of what it is like to be an ordinary white North American, English or Irish person, on holiday from a world supposedly threatened by the encroachment of cultural and racial difference.[5] The encounter with Aboriginality in the travel narratives of the 1990s, like those by Bryson, Caulfield and McCrum, question such myths and reveal uncomfortable aspects of contemporary Australian life for national and international audiences.

The Myths of Ordinary Australia

Non-Indigenous Australians frequently complain that foreigners have skewed perceptions of their character, land and heritage, and that they are more often than not the victims of misrepresentation.[6] This complaint, however, ignores the roles Australian institutions, such as the tourism industry, perform in the production and dissemination of representations of Australian life. During the last twenty years tourism has become an increasingly significant factor in the domestic economy and society.[7] Australia is highly dependent upon income derived from international tourists, and the growth of this industry in Australia has been nothing short of phenomenal. The success of tourism in Australia is attributable to formal marketing campaigns, as well as the international success of Australian popular cultural exports. In the latter case, during the 1980s and 1990s a range of media products defined Australia for foreign audiences through a variety of appealing landscapes and cultures. Examples of such products include internationally successful films like *Crocodile Dundee* (1986), *The Adventures of Priscilla, Queen of the Desert* (1995) and *Strictly Ballroom* (1993); television shows like *Neighbours, Home and Away, The Secret Life of Us,* and the late Steve Irwin's *The Crocodile Hunter;* popular music produced by bands such as Savage Garden, INXS, Crowded House, Midnight Oil, Powderfinger; as well as Aboriginal art and cultural products in a range of guises, from Western Desert acrylic paintings to David Gulpilill's performance in *Crocodile Dundee* to Tracey Moffatt's photography.

In Australian tourism discourse and popular culture since the mid-1980s two dominant paradigms have influenced the production of images that define Australia's attractiveness to foreign travellers: the *exotic* and the *ordinary*. *Exotic Australia* projects images of rugged and wild landscapes (the 'Outback,' Uluru/Ayers Rock, the Great Barrier Reef, rainforests, deserts), and showcases a strange and ancient Aboriginal culture, and a hardy white settler heritage. Alternatively, *ordinary Australia* represents a nation of relaxed, middle-class urban lifestyles and accommodating rural landscapes, of friendly, informal, open and down-to-earth people. This is an 'ordinary' place that Bryson (2000) describes as "nice," "stable and peaceful and good" (4), "comfortable and clean and familiar" (10), inhabited predominantly by white people of "modest yearnings for respectability without fuss" (98). These paradigms – the exotic and the ordinary – complement each other by presenting Australia as amenable to a diversity of travel tastes and styles.

Across different texts and sites one paradigm will tend to dominate or subsume the other. In much Australian tourism discourse, as well as in many travel narratives like those discussed here, the ordinary dominates the exotic such that readers and visitors are seemingly offered representations of a place that affords travellers adventure and a sense of estrangement, while assuring them that the experience of everyday Australian life and the ordinary aspects of travel – unhindered movement, familiar shopping and business environments, safe pedestrian experiences, a common linguistic and cultural heritage – will facilitate the pleasures of the visit. Indeed many recent travel texts about Australia parody or satirize the images and stereotypes associated with exotic Australia. Mocking the exotic is a way of un-settling the unreasonable expectations of the writers' own fellow countrymen and potential readers, while at the same time reinforcing the idea that the essential pleasures of travel through Australia lie in its 'ordinariness.'[8]

Yet the treatment of the ordinary in travel texts like those of Bryson, McCrum and Caulfield is ambivalent. Understanding how ordinariness functions in these texts, and in contemporary travel writing generally, is assisted by considering its functions in Australian popular culture. Some schools of criticism consider popular culture as inherently concerned with the activities and interests of ordinary people (see for instance Gans 1999), and indeed figure the ordinary as a privileged sphere in which "utopian and political aspirations … crystallize" (Kaplan and Ross 1987: 3). Yet while it is equally important not to consider the quotidian as a priori conservative, criticism should avoid any utopian essentializing of this concept. 'Ordinariness' is a trope that functions within popular culture to achieve various and competing ideological objectives. This becomes clear when one considers the politics and aesthetics of ordinariness in Australian popular culture.

For Meaghan Morris ordinariness is a "sacred, secular value" in Australian popular culture (Morris 1998: 108). David Carter notes that images of ordinariness in Australian film, television and print have long been one of the most pervasive means through which the discourse of egalitarianism circulates in Australia (Carter 2006: 360). Ordinariness manifests as a range of aesthetic codes frequently characterized by informality in dress, appearance, language and attitudes, including, on the one hand, distaste for and resistance to acts of deference towards class or status, and on the other hand concern for the significance of everyday events and lives. For example, Tom O'Regan observes that Australian films frequently "focus on people who would be in the periphery [of Hollywood films], and cast physical types into central roles who would normally be cast into supporting roles" (O'Regan 1996: 245). O'Regan suggests that the exploration of the everyday and celebration of the ordinary in Australian film – and I suggest popular culture in general – represents an ideological commitment to the "human scale" and "truth to the actual" (245).

Ordinariness functions, then, in Australian popular culture as a set of signifiers of popular democracy – a confirmation of 'Australian' values of egalitarianism, the 'fair go,' mateship and anti-elitism. Moreover, ordinariness performs the role of mediating and managing dialogue and discourse between otherwise distinctive and potentially conflictual social groups. As political scientists Judith Brett and Anthony Moran note, "the term 'ordinary' comes up again and again when Australians talk about their social world and their place in it, as do synonyms, 'average,' 'normal' and 'everyday'" (Brett and Moran 2006: 2). During the 1990s terms like 'ordinary people' and 'ordinary Australians' were often used in the public sphere in a pejorative manner that reflected white middle-class cultural values and mores in opposition to the transformations in class, gender and – most significantly – race relations that were affecting the body politic (1–3; see also Hage 1998; Sinclair 2004; Frow 2007).

The 1990s presented Australia with many significant economic, social and legal challenges that disturbed popularly held assumptions about history and race relations. Of most significance for present purposes were the challenges to the political and moral authority of white Australians that came through the legal and political struggles by and on behalf of Aboriginal people.[9] Associated with this were the efforts of many to counter long-held narratives about the history of European 'settlement,' and to reveal the extreme social, economic, legal and health inequities experienced by most Aboriginal people. Against such efforts the rhetoric of 'ordinary Australia' was frequently deployed by conservative political constituencies to confirm the moral legitimacy and authority of white Australia. Bryson, McCrum and Caulfield were travelling through and writing about Australia at a particularly intriguing time in the nation's history, as well as at a moment when the politics of the ordinary were deeply contested.[10]

In these travellers' narratives the ordinariness of Australia is represented in two ways. First, ordinary Australia is experienced as an *ethnicized* and stereotypical form of whiteness that for transnational Anglo-Celtic publics speaks to family resemblances with white Australian culture. That is, the texts exploit and critique the myth projected through Australian popular culture that white Australians are more or less just like their English, Irish and (white) North American 'distant cousins,' who in turn are offered the promise of enjoying the experience of finding and being their true 'ordinary' selves in Australia. At the same time, the texts unsettle the idea of the ordinary per se – particularly the ordinary as a utopian and transformative element within social life. This arises from the observation that, in the race debates that dominated Australian public life in the 1990s, 'ordinariness' was frequently mobilized by conservative white Australians to deflect accusations of racism and to deny their complicity with the history of European colonization in Australia and the devastating consequences this has had for Aboriginal Australians. While white Australia solicits visitors with the promise of a redeemed sense of an ordinary (customary, normal) Englishness, Irishness or American-ness, for Bryson, Caulfield and McCrum the encounter with Aboriginality potentially disrupts such fantasies, as these travel writers discover that Aboriginality can, seemingly, never be ordinary in the context of white Australia.

Encounters with Ordinary Australia

A common cliché employed in many travel narratives about Australia is that the nation is known abroad only by its clichés. Even in the late 1990s British and U.S. travel writers still employed the conceit of Australian anonymity: Australia is either completely unknown or only known through the stereotypes of popular culture. For Caulfield, Australia is a "vanishing point" (9): "we know real people go there and come from there, but what goes on when they are there is sketchy" (Ibid.). Likewise Bryson reminds his North American readers that "we pay shamefully scant attention to our dear cousins Down Under" (4). Alternatively, McCrum admits his prior knowledge of Australia amounted to:

> a fantastically ill-considered cocktail of Castlemaine and Fosters ads, with stray bits of *Crocodile Dundee,* Dame Edna, *Strictly Ballroom* and *Neighbours* thrown in. It was a land of vast, dry, empty spaces where rough, suntanned rednecks drove down dirt roads in pick-up trucks piled high with tinnies of amber nectar, occasionally stopping to grunt appalling, chauvinistic remarks ... (xv)

The conceit of Australian anonymity defines a readership and addresses the stereotypes of everyday Australian life that circulate within the writers' home cultures. Rejecting such stereotypes, each narrator informs their reader that they intend to discover the "real Australia" (Bryson 2000: 11), "an Australia that was far away from the clichés" (Caulfield 1999: 42) and "a more sophisticated picture" (McCrum 1997: xvii) of Australian life. At the same time, the conceit of Australian anonymity addresses the reader in a manner that assumes identification with a transnational interpretive community with a common linguistic and ethnic – Anglo-Celtic – heritage ('our dear cousins Down Under').

In looking for the 'real Australia,' Bryson discovers that this country is the *secret* landscape of the lost childhood/adolescence of the U.S. baby boomer. This is an ordinary Australia free of the anxieties of the contemporary world; a place that affords the innocent pleasures of pre-affluent capitalism, and is instantly recognizable and consumable because it is so North American. Two commodities define this fantasy of Australia as North American for Bryson: potentially unhindered movement and actual nostalgia. For instance, driving through the rural landscape west of the Blue Mountains that ring Sydney, Bryson discovers a "delightful and accommodating" (76) countryside that is "the American Midwest of long ago" (77). He comes to a startling realisation: "*I was driving through my childhood*" (77; *emphasis added*). In the realm of ordinary Australia, Bryson finds himself in America: a comfortable white 1950s mid-West and, in Graeme Turner's words, a "highly idealised deeply nostalgic vision of America" (Turner 1994: 115).

The experience of Australian space as American facilitates for Bryson, and by extension the implied reader, a sense of being unburdened by the anxieties of the times. Australia is so big and sparsely populated that one can immerse oneself in an escapist fantasy from the claustrophobia of modern life. If, in secular modernity, hell is other people, then the spaciousness of Australia, according to Bryson, is heaven on earth: one can travel vast distances without encountering others. The troping of Australian space as North American is also associated with what is perceived as Australia's lack of distinction, which in turn frees the writer/tourist from having to say anything original about the place. It also frees Bryson from feeling the need to consult anyone who might have some degree of expertise on Australian life, politics or culture: Although Bryson meets and converses with a number of 'ordinary' Australians, the encounters are relatively pedestrian, and nowhere does this internationally renowned author describe any meetings or conversations with anyone of note. While Bryson constantly reminds readers of how little he and other U.S. citizens know about Australia, it is clear that there is no need for introduction, translation, explanation or guidance from locals, because apparently Australia announces itself. For Bryson, as Morris has observed for other North American travellers from Jesse Ackermann (1857?–1951) on, "the one distinctive feature of Australian culture is its *positive unoriginality*" (Morris 1988: 245). In its rhetorical performance of the ordinary U.S. tourist, Bryson's text suggests that Australia is no more than a banalized (Baudrillard 1990) version of the United States (and all the more enjoyable for it), where the jaded white Anglo-

American can experience the simple pleasures of life – sunshine, driving, getting away from other people, and so on. Nonetheless, as discussed below, it is a performance that is ironized for the reader by, amongst other things, Bryson's encounters with Aboriginality.

If *space* is the coordinate that Bryson celebrates in his rhapsodizing of ordinary Australia, then for Caulfield it is *heritage* that functions as one of the most significant vectors for her ambivalent – and at times distressful – experiences of Australia. In part this is due to the family connection, as tenuous as this might be, that Caulfield has with Australia. Caulfield travels Down Under to find what happened to her 'Uncle Caulfield,' who according to family legend acquired fame and fortune in Australia before mysteriously disappearing. As with most Irish family tales the truth about Uncle Caulfield turns out to be far less romantic than the myth, and when no trace can be found of her uncle, Caulfield decides to make of her journey an exploration of "what became of the Irish" in Australia (Caulfield 1999: 41).

It is a journey that involves a range of styles and encounters with black and white Australians, and one that particularly relies on mediated experiences such as those afforded by museums and heritage centres. It is in these establishments that Caulfield glimpses the transformative possibilities of ordinary Australian culture. Of the Museum of Sydney, Caulfield states:

> London museums are kings, queens, geniuses, the fantastic and best of the world; Sydney collected fragments from lost people, brought here to disappear. But these people built the country. Ordinary, forgotten people built European nations but no importance is attached to them, there's no regret for them. In Australia, the almost sentimental but actually quite touching mourning for these ordinary people that time forgot, lets you feel how a democracy has to be at heart. Every hand that touched the land, counts for the same. (56)

It is in part the myth of Australia as a place where 'ordinary' people – and in particular ordinary Irish people – matter that attracted Uncle Caulfield to Australia in the first place, and likewise attracts his great-niece. Australia has been deeply influenced by its Celtic (especially Irish) heritage, and the idea that Australia might be a site in which the 'myth of Irish dispossession' can be redeemed has a strong influence in the national culture. But while Caulfield is on the lookout for signs of distant kinship and the Irish-Australian legacy, she is deeply suspicious of the way many white Australians embrace their Irish heritage. For Caulfield the 'heritage industry' represents an "addiction" and "booming cottage industry spreading into mainstream Australian literature" (214) that justifies white colonization and threatens, if only through sheer volume, to silence any black voices from the narrative of the nation's history. Dominated by "[h]eartrending tales of hard-done-by Anglo-Celtic convicts and immigrants" (214), the heritage industry became "the white land rights claim" (214–15) at a time when Aboriginal voices were demanding equality, territorial rights, political sovereignty and compensation for past criminal actions by the State and its representatives. Heritage for Caulfield functions as a symbolic system reflecting her own ambivalent identification with her Irishness and her engagement with Australia, while remaining a troubling aspect of white Australian culture when it elides the historical realities of the colonial past.

McCrum, too, encounters heritage during his journey through Australia, but such experiences are set within the context of more general questions about the importance of class and race in Australia and how these contrast with English society. Against the degraded stereotypes of Australian popular culture, McCrum undertakes the task of discovering a supposedly more accurate picture of Australian life. In doing so he follows a long line of middle-class English travellers who have explored white, Anglo society in Australia (from Anthony Trollope to Elspeth Huxley and Geoffrey Moorhouse). This inevitably carries with it comparisons between Australia and the 'mother' land, especially in relation to class.

McCrum reflects that he formerly held a fairly stereotypical English attitude towards the antipodes: "For the English, Australia and Australians were still – let's face it – a bit of a joke. Just the accent made us laugh. All the more so if it had any pretensions to anything" (McCrum 1997: xvi). Australia is a place of debased 'English' language, intelligence and morality. In this regard McCrum is playing on a significant myth that dominates the relationship between Australia and England: the transformative nature of the Australian experience (positive or negative) on the English subject.

Yet while McCrum begins by reiterating the negative stereotypes about Australia that commonly circulate within English domestic culture, his journey leads, predictably, to opposite conclusions. Australia for McCrum, despite certain limitations, is overwhelmingly a land of promise, one that remains inviting to English tourists as well as migrants. Indeed British – and specifically English – travel writing on Australia has always addressed domestic readers as potential migrants, given the special historical relations between both nations. In this respect, McCrum's narrative reflects what Mark Gibson suggests is a primary attraction of Australia for English immigrants (be they labourers or expatriate cultural studies academics): namely, that it is "a stage for a certain transcendence of Englishness" (Gibson 2001: 280). Such transcendence, by this reasoning, is not necessarily a rejection of English identity, but rather a cultivation of those values of Englishness that are most notably free of the limits imposed by a highly structured domestic class system. If past English travellers emphasized the potential of Australia as a workingman's paradise, contemporary Australia offers the kind of relaxed egalitarian middle-class lifestyle that is apparently beyond the reach of many in England itself.

The transformative potential of Australia on the English subject is highly ambiguous in *No Worries* and appears to be more a case of stylistic re-evaluation than politically substantive. This conclusion is suggested by an anecdote that McCrum gives in the preface of his book. Upon entering a restaurant in his parents' London club he is informed that other patrons disapprove of his lack of a tie. The scene contrasts McCrum's souveniring of Australian informality with the stereotypical stuffiness of upper middle-class Britain: "[p]eering through the panelled gloom, I saw a scowl above a fat, pinstriped belly and sincerely wished I had never got on that returning plane" (xii). The scene resonates with McCrum's attitude toward Australian culture and seems to reflect Ian Baucom's insight into the influence of the colonial experience on English identity: for modern travellers, as for those past, "[t]he empire ... is less a place where England exerts control than the place onto which the island kingdom arrogantly displaces itself and from which a puzzled England returns as a stranger to itself" (Baucom 1999: 3). In McCrum's case, however, the estrangement does not appear to run too deep. If anything, the anecdote reinforces the impression that McCrum finds in Australia a place and culture that allows him the freedom to be just who he is: a young middle-class English male. And indeed in his encounters with sophisticated urban middle-class whites from gay households in Sydney to yuppie swingers in Brisbane, to the modern version of the rural 'squattocracy,' McCrum explores the different bourgeois lifestyles that (for some) define the most attractive aspects of ordinary Australian life.

Yet as the narrative progresses his encounters with whites increasingly become inflected by discourses on race and Aboriginality, and the pervasiveness of debates about the definition of Aboriginality – who are, and who may speak for, Indigenous Australians – in everyday engagements with white Australians leads McCrum to seek more intimate contacts with Aboriginal people. For McCrum, Australia is a successful modern multicultural melting-pot, open to white English explorers to wander through, consume and enjoy at their leisure. This is consistent with Bryson's take on the nation, but contrasts with Caulfield's unnerving experiences of everyday misogyny and racism. Nevertheless, although McCrum and Bryson initially exploit the notion of Australia as a space in which 'difference isn't inherently threatening,' like Caulfield they discover that this myth is especially untenable when considered in light of their experiences of Aboriginality.

Robert Clarke

The Encounter with Aboriginality

During the 1980s and 90s Indigenous culture became an increasingly prominent signifier of Australia abroad. The perverse contrast between the valuing of 'traditional' Aboriginality in the context of promoting the Australian nation and the material conditions experienced by most Indigenous Australians is not lost on visitors like Caulfield, McCrum and Bryson, who provide their readers with various details and statistics demonstrating the degree of deprivation that many Aboriginal people experience in this country.

One intriguing aspect of Bryson's *Down Under* is the relative absence of many of the signs of Aboriginality that are such a significant resource for other travel writers at this time. As Bryson informs his readers, Aborigines are "Australia's forgotten people" (192) and the "world's invisible people" (196) – and his text might appear to confirm these statements in its relative neglect of Aboriginal subjects. But Bryson is not such a blind traveller that he could have journeyed through Australia in the 1990s and not noticed how Aboriginality was informing and shaping public discourse. Bryson's primary strategy is to comment on white perceptions of Aboriginality. Indeed Bryson's first encounter with Aboriginality is through racist remarks proffered by white Australian tourists about the 'Aboriginal problem,' including those of a retired Canberra solicitor:

> 'They want hanging, every one of them.'
> I looked at him, startled, and found a face on the edge of fury.
> 'Every bloody one of them,' he said jowls trembling ...
> Aborigines, I reflected, were something I would have to look into. But for the moment I decided to keep the conversation to simple matters – weather, scenery, popular show tunes – until I had a better grasp of things. (25)

When Bryson decides to engage with Aboriginal Australia he chooses as his first destination Myall Creek, the site of the massacre in June 1838 of 28 unarmed Wirrayaraay men, women and children by white settlers. The site represents one of the few instances from the nineteenth century when white murderers of Aborigines were tried, convicted and punished for their crimes. Yet when Bryson arrives at Myall Creek there is apparently nothing "to indicate that here ... was where one of the most infamous events in Australian history took place" (200). Even in the nearby town of Bingara the tourist information centre possesses no history of the site. Bryson is told that there is no interest in recording what happened to the local Indigenous people; the event was unremarkable as it was only one of many massacres that took place in the region. A local journalist states, "It's kind of ironic when you think about it. Myall Creek's not famous for what happened to the blacks here, but for what happened to the whites. Anyway, you wouldn't be able to move in this country for memorials if you tried to acknowledge them all" (203). When Bryson asks the journalist whether there are any Aborigines still living in the area, he is told, "Oh, no. They're long gone from round here" (Ibid.).

Bryson's comments need to be treated with circumspection, and are perhaps consistent with his customary egocentric travel persona. During the 1990s a memorial at Myall Creek was organized. This brought together Indigenous locals (including descendents of survivors of the massacre) as well as non-Indigenous residents. This organization would likely have been underway when Bryson visited the town. Irrespective of whether this oversight was deliberate, it reinforces the sense that for the kind of tourist that Bryson supposedly represents, Aboriginality in the form of Aboriginal people is absent from everyday, 'ordinary' Australian spaces, although Aboriginality understood as a set of discourses on and about Aboriginal people is pervasive. At the same time, Bryson's narrative clearly articulates a point frequently made by a number of prominent Australian historians: Despite the high values placed on the memorialisation of Australians' involvements in international conflicts, until very recently no official commemorative sites have recognized the frontier wars on Australian soil between Europeans and Aborigines (see for example Inglis 2001).

For Bryson the absence of Aboriginal people in public life stands in contrast to the prominence of speech – frequently racist – about them by whites, as evidenced by the anecdote of the racist solicitor. Moreover, when he does see Aboriginal and non-Aboriginal Australians together they appear to exist in separate spheres. Observing the black and white pedestrians in Alice Springs's Todd Street Mall, Bryson notes how the "two races seemed to inhabit separate but parallel universes. I felt as if I was the only person who could see both groups at once. It was very strange" (279–80). Reflecting on this he notes the general absence of Aboriginal people from everyday Australia:

> Above all, what is perhaps oddest to the outsider is that Aborigines just aren't there... you wouldn't expect to see them in vast numbers anyway, but you would expect to see them sometimes – working in a bank, delivering mail, writing parking tickets, fixing a telephone line, participating in some productive capacity in the normal workaday world. I never have; not once. (283)[11]

As with his depiction of Myall Creek, Bryson's statement rehearses the myth of a 'lost' or vanished Indigenous subject. Moreover, it plays into stereotypical representations of Aboriginal people in terms of physical attributes that many who identify as Indigenous do not necessarily possess or consider integral to their identities. On the other hand, when read as a self-conscious performance, Bryson's 'blindness' to Aboriginality might be considered another self-parody that undermines the legitimacy of the experiences that Australia is supposed to offer. Concluding that solutions to the 'problems' of Aboriginal Australia elude him, Bryson writes, "I just sat for some minutes and watched these poor disconnected people shuffle past. Then I did what most white Australians do. I read my paper and drank my coffee and didn't see them anymore" (284).

While it is possible to interpret this scene as demonstrating, as Debbie Lisle does, "the reluctance of conventional travel writers to engage with difficult political questions" (Lisle 2006: 176), it is equally possible to read the passage as a form of auto-critique. Bryson's narrative suggests that the fantasy of ordinariness for the white tourist, as for white Australians, requires a kind of selective blindness in matters relating to Aboriginality. Only by not seeing the realities of the racial divide between Indigenous and non-Indigenous Australians, the text suggests, can the tourist actually enjoy the pleasures that ordinary Australia affords.

By contrast, a great deal of the angst that Caulfield expresses in her text comes from being unable to 'not see' the unjust treatment of Aboriginal people that permeates everyday Australian life. As she travels around Australia she is constantly shocked and repelled by the white racism she observes, and complains that she "couldn't stand being welcomed and helped by people who always turned out to be racist bigots" (Caulfield 1999: 308). Her search to find out "what became of the Irish" (41) in Australia is given a personal twist when Caulfield discovers a story about a lagoon that bears her family name. Also known as Murdering Lagoon, and located near Cooktown in far North Queensland, the author learns how in November 1873 "a gang of government officials and gold miners hunted between eighty and 150 Aboriginals.... They shot them all and left the bodies floating in a lagoon. There was an enquiry, but the slaughter was held to be a reasonable act of self-defence" (42). Caulfield remarks that "as the fantasy of the Caulfields in Australia evaporated I was left with the cruel fact of Caulfield's Lagoon. Whatever the lagoon's name had to do with my family, I just sensed that following the trail of this story would lead to an Australia that was far away from clichés" (42).

These clichés include the usual tourist images of the 'bronzed Aussie' beachgoer, the stoic Outback stockman, the Barrier Reef, and, for Caulfield, the mythologies of the Irish in Australia. In her examinations of various sites and places she is sensitive to any signs of the conscription of Irishness within narratives of Australian identity, and particularly wary of any form of sentimentalization of Australian Irish heritage. In this regard she seems to be responding to a phenomenon that Jennifer Rutherford has described as the "Irish conceit" (Rutherford 1998). Rutherford argues that during the 1990s 'Ireland' was deployed in Australian nationalist movements such as those on the political right wing including Pauline Hanson's One Nation Party. Rutherford claims that the rhetoric of this conceit subsumes Aboriginal histories of dispossession within a mythology of "Irish suffering, dispossession and the tragic impossibility of any form of redress for the Australian victims of the Irish diaspora" (197).

Such sentimentalization of Irishness is rendered suspect in Caulfield's text. For Caulfield, the Irish Australian heritage is most poignantly evident in relation to Aboriginality. As she notes ruefully, this is in part due to the closeness of Irish stockmen and settlers and Aboriginal communities: "There is a lot of Irish blood from lonely white stockmen who'd taken indigenous wives; a lot from 200 years of rape" (130). On the other hand, there is the legacy of the work of Irish missionaries and religious orders including the Christian Brothers, and the effects they have had as agents of colonialism. And there is the legacy of Caulfield's Lagoon. Searching for the lagoon in the company of Billy, an Aboriginal elder, and a white guide, Phil, Caulfield is appalled by the insulting manner in which Billy is treated by various whites during the course of a daytrip; unavoidably for Caulfield, the legacies of colonialism are a disturbing and pervasive influence on everyday life in Australia.

Likewise, McCrum's travels constantly bring him face to face with the politics of race in Australia, and more particularly with the issue of Aboriginality. McCrum casts a keen eye over those 'sympathetic' whites, from those who inevitably express standard racist views to those who attempt to engage with Indigenous Australians in ways that encourage reconciliation between the communities. In the characters of Mrs. Donovan and Don Green, McCrum encounters the kind of right-wing rhetoric about Aboriginality that would become a feature of the One Nation Party's ideology in the middle of the decade. For instance, Mrs. Donovan, a white woman who manages an inner-city boarding house initially expounds her anti-racism but then explains to McCrum that Aboriginal people, for her, are just too different: a race of no-hopers who can't "get their act together," pre-disposed to alcoholism, and anyway most of them "are only part Aborigine" (McCrum 1997: 63). The accusation that most Aboriginal people are both incompetent and inauthentic as a result of mixed heritage speaks to long-held white anxieties and ideologies concerning miscegenation as well as to the legitimacy of Aboriginal cultural and territorial claims. Don Green from Kalgoorlie is another 'sympathetic' white whose arguments about the 'purity' of Aboriginal blood reflect covert racism. When McCrum questions Green's views on 'half-caste' Aborigines and suggests that such ideas reflect a non-Indigenous strategy for denying Indigenous people rights and compensation, Green objects and claims that it is Aborigines who are in fact exploiting whites for sympathy and money in what amounts to a threat to the nation's sovereignty: "We've got to be Australian first, and whatever else we are second" (223). In Mrs. Donovan's and Green's ideology Aborigines are not simply undeserving; they are also disadvantaging white Australians. They express what Steve Mickler has termed the 'myth of privilege': the idea that affirmative action, territorial reclamation and compensation efforts by and on behalf of Indigenous people – the most disadvantaged group in Australia – amount to 'special treatment' and have established Aboriginal people as "an advantaged class within, and at the same time apart from, Australian society as a whole" (13).

On the other hand, McCrum encounters other whites who attempt to engage with Indigenous Australians through a critique of white ideologies: people like Ian Greenwood, a white man initiated by the Banggala people of South Australia, and 'Philip' a priest from the Lutheran Mission in Alice Springs, who introduce McCrum to the complexities of 'traditional' Aboriginal lifestyles and beliefs and their meaning within a modern context. Both men embody the possibilities of cross-cultural dialogue. So does well known historian Cassandra Pybus, who exemplifies for McCrum the spirit of a small group of white Australians who are "actively questioning their right to be on their own land; who come with thoughts like, 'I still don't know what I have to do to act responsibly as someone who appreciates that they're the inheritor of stolen property'" (279). Pybus' book *Community of Thieves* is one of many written by white intellectuals in the last twenty years that explore the meaning of white postcoloniality, but as the controversy around Pybus' book demonstrates to McCrum, expressions of white sympathy are not guaranteed an appreciative response from Indigenous people themselves.

Contrasting with these white experiences of Aboriginality are McCrum's encounters with a number of Indigenous people. McCrum's meetings with May O'Brien and Irene Stanton, for example, speak both of the distressing history of racial dispossession as well as the pride with which many contemporary Aboriginal people embrace their identities. Answering the racist discourses of Mrs. Donovan and Don Green, O'Brien and Stanton situate themselves as Aboriginal *and* Australian. Stanton argues that white racism denies her acceptance within the white community: "For me my stance is very simply: I got brown skin, doesn't look as if I'm ever going to be accepted totally by the non-Aboriginal community, because of the colour of my skin and because my heart's with the Aboriginal people, and our struggle isn't over, and I've got to be part of it" (200). Yet the fact of her non-acceptance by the white community is not necessarily a hindrance to her acceptance of a national identity, and in making such a claim Stanton refuses the racialization of Australian identity and gestures towards the necessarily hybrid nature of the contemporary nation.

Stanton's comments recognize the ambivalent status of Aboriginality in everyday, 'ordinary' Australian life. Cultural Studies scholars Alan McKee and Steve Mickler make similar observations. They argue that white Australia constructs Aboriginality as fundamentally – indeed fatally, in Baudrillard's sense – different. McKee argues that whenever Aboriginal people are represented through the signifiers of the ordinary, the everyday and the banal they are immediately rendered inauthentic – as not truly Aboriginal (198–99). Mickler argues that the 'extra-ordinariness' of Aboriginal people is a result of the lack of a formal recognition of their sovereignty. It is a condition, therefore, of white (post)coloniality that Aboriginality is always conditioned by *exceptionality*. White Australia thus denies Aboriginal Australians 'the right to be ordinary' because their sovereignty has not been fully and properly recognized. Mickler argues that in Australia,

> racial prejudice consists partly of a prejudice towards the idea of social groups being 'exceptions to the rules' or exceptionalism in general.... [P]opular prejudice is tied up with a popular democratic hostility to cultural exception when this exception is thought to be ... a bid to be free of the universal constraints of governance. (Mickler 1998: 305)

The insights of McKee and Mickler, and the ideas expressed by Stanton, do not presume a static model of identity. And indeed Stanton's comments, like those of other Aboriginal people in McCrum's book, attest to the determination of many within Aboriginal and non-Aboriginal Australia to resist the everyday prejudices of Australian culture and to transform the ordinary culture of the nation in a manner that accepts a multiplicity of identities.

Conclusion

Insofar as the three travel texts discussed here engage with the myths of ordinary Australia, they map the effects of a dissonance experienced by many white liberal travellers when everyday encounters with Indigenous people and their culture are sought. In the cultural economy of Australian tourism, Aboriginal people and their cultures are called upon to serve particular ideological functions that confirm or at least do not disturb white hegemony. When Aboriginal people do not conform to such stereotypes they are either considered inauthentic and rendered invisible, or positioned in terms of negative stereotypes of cultural and racial degeneracy that reinforce assumptions of Indigenous exceptionality (for example the supposed inability of Aboriginal people to cope with civilization, their 'natural' violence and 'social disorderliness').

However, for travellers like Bryson, Caulfield and McCrum, the denial of the capacity of Indigenous Australians to be ordinary in modern Australia serves to disturb the claims of ordinariness by white Australians, and consequently to disrupt the notion of Australia as a comfortable site for the international visitor to experience a sense of ordinariness. This is because the tourist's embrace of the white ethnicities that Australian ordinariness supposedly promises is contingent upon the acceptance of an ideology that renders Aboriginal people 'exceptional' and denies them the right to be ordinary in their own country and communities. In their ambivalent ways, the books by Bryson, Caulfield and McCrum demonstrate how trips through the everyday and ordinary spaces of postcolonial cultures like Australia can have estranging effects.

While 'ordinary Australia' is a principle co-ordinate for the travels of Bryson, McCrum and Caulfield, it is also a target for their critiques of aspects of Australian life. Their engagements with everyday life in Australia, with Aboriginality and with Aboriginal subject serves to highlight the racism prevalent in many social situations and actors. Yet their travels also bring them – except for Bryson – into contact with Aboriginal people, some of who work within the travel and tourism industries. As attested by publications like Paul Kauffman's *Travelling Aboriginal Australia: Discovery and Reconciliation* (2000) and the Lonely Planet's *Aboriginal Australia & the Torres Strait Islands: Guide to Indigenous Australia* (2002) the number, variety and sophistication of Aboriginal owned and/or managed tourism operations developed rapidly during the 1990s.[12] While serving clear economic imperatives, these initiatives can also serve progressive social and cultural functions. Caulfield captures the significance of such innovations in her final tourism encounter in Broome, Western Australia. She is guided around the town by Ben, a local Indigenous man who had once played the lead role in the popular musical *Bran Nue Dae*, the main theme of which Caulfield explains is that "there's no Australian who isn't a couple of degrees away from Aboriginal blood – I mean in their veins or on their hands" (291). The other tourists on the walk include a family of two adults and their three boys. As they walk around Broome, the parents display a sympathetic concern for understanding the heritage of the place, Indigenous as well as non-Indigenous. Ben describes the relationships between the cultures that make up Broome's famous cultural melting pot, and inevitably the legacies of Irish Catholic influence on local Aboriginal culture. Ben tells them, "I grew up knowing all about the history of Ireland and Germany, nothing about Aboriginal history. I knew Dublin all right" (296). When the young boys start misbehaving, Caulfield describes how the mother "suddenly grabbed the two nearest and turned them to face Ben" and then tells them, "'Listen to this. We don't know about all this. You live in a very privileged world and you don't know about this'" (293). The scene serves as a parable to end Caulfield's strange search for her own Australian family connections. It provides a hopeful image of the tourist encounter between Indigenous and non-Indigenous people as an opportunity for addressing and redressing the myths and misconceptions about identity, culture and history that still permeate everyday discourse in Australia. And it demonstrates the manner in which Aboriginal agency is influencing how ordinary Australia is understood and represented by Australians and the rest of the world.

Robert Clarke is a senior lecturer and Head of Discipline in English in the School of Humanities, University of Tasmania. He is the author of *Travel Writing from Black Australia* (Routledge, 2016), and the editor of *The Cambridge Companion to Postcolonial Travel Writing* (Cambridge University Press, 2018) and *Celebrity Colonialism: Fame, Representation and Power in Colonial and Postcolonial Cultures* (Cambridge Scholars, 2009).

Notes

An early version of this article appeared in the Annual Yearbook of the Department of English, University of Delhi, 2004. I wish to thank Helen Gilbert and Rimli Bhattacharya for their generous advice on early drafts of this paper.

1. It is convention in Australia to spell indigenous with a capital 'I' when referring to Indigenous Australians (that is people of Aboriginal and Torres Strait Islander heritage), and to use lower case when applying the term generally, and I have followed this convention throughout the paper.
2. The report claims that 1 in 3 children were removed from their families. This has been hotly contested. As historian Robert Manne points outs, it is almost certainly wrong. Manne claims the "lower estimate of one in ten is far more soundly based" (25). This still represents a very large number of children removed from their families – Manne estimates 20–25,000 since 1901 (27).
3. An apology to members of the Stolen Generations was formally made in the federal parliament by Prime Minister Kevin Rudd on 13 February 2008 after taking office from John Howard who had governed Australia for 11 years and had vigorously resisted calls for such an apology for most of that time.
4. Other titles by visiting authors that could be examined from the perspectives presented here include Tony Horwitz's *One for the Road* (1987), Howard Jacobson's *In the Land of Oz* (1988), Brian Johnston's *Into the Never-Never* (1997), Jan Morris' *Sydney* (1992), and Geoffrey Moorhouse's *Sydney* (1999).
5. In this paper I will use the term 'white Australia' to refer to that section of the Australian community that identifies its European heritage with Great Britain and Ireland. The texts under discussion here clearly demonstrate their narrators' identification with an Anglo-Celtic heritage.
6. One finds an expression of this anxiety in the reviews of Bryson's *Down Under* that appeared in the Australian press. See for instance Dale (2000).
7. Faced with a declining economy in the early 1980s, the Australian government invested heavily in a campaign to energize the tourism industry. Until the late 1970s, given Australia's remoteness and the high costs involved in travelling there, international travellers (that is, short-term leisure visitors) had been an infrequent commodity. Since the implementation of the Federal Government's tourism campaign from 1984 onwards, the

annual number of international tourists has grown rapidly. Tourism numbers doubled over the 1990s, and by the end of the decade the industry accounted for 6 per cent of Australia's GDP, 15 per cent of total export earnings, and employed approximately 700,000 people (8 per cent of the workforce) (see Craik 2001: 96).
8. Bryson, McCrum and Caulfield each engage in parodies of the popular cultural stereotypes and touristic clichés associated with the Australian exotic from the 'Bronzed Aussie' beach culture to the 'Crocodile Dundee' culture of the Outback. Many other travel authors play with these cultural stereotypes from Howard Jacobson in *In the Land of Oz* (1988) to Peter Ruehl in *American Downunder* (1992) to Tony Horwitz in *Into the Blue* (2002) (see also Note 3 above). On the other hand, many recent travel books on Australia are deeply invested in the exotic imagery of the nation, especially when it comes to representations of Aboriginality. For example, the raft of 'New Age' travel books that emerged in the late 1980s and 1990s with titles like *Mutuant Message Down Under, Two Men Dreaming*, and *When You See the Emu in the Sky* exploit the spiritual and mystical values associated with 'traditional' and 'ancient' Aboriginal spirituality and cosmology, for a very large international audience (see Clarke 2009 and 2016).
9. It is beyond the scope of this paper to provide a full history of these movements and changes. For a valuable introduction to this history interested readers should consult Broome (2001).
10. It should also be noted that the books under discussion were released in the lead-up to the Sydney 2000 Olympic Games, and were amongst a large number of travel books published about Australia by foreign visitors around this time.
11. Intriguingly, Bryson's account here contrasts with those of other travelers who do see Aboriginal people in the roles that Bryson describes. For example, Sven Linquist states: "Aborigines are to be found working in private and government offices, as shop assistants, cleaners and parking attendants, and as troublesome layabouts in the parks. I see them as clients at court, as hospital patients, as artists in art galleries and occasionally as restaurant guests..." (44). Admittedly, Lindquist's description relates to his experiences in Alice Springs, which has a large Aboriginal population. Yet, for Lindquist as for Bryson, the most salient feature of the presence of Aboriginal people is the apparent absence of civil relations between Aboriginal and white Australian subjects, and the lack of opportunities for intercultural discourse.
12. The *Black Pages* website, http://aboriginaltouroperators.com.au/blackpages/blackpages.htm, provides a long list of Aboriginal and joint venture companies specializing in Aboriginal cultural tourism.

References

Baucom, Ian. 1999. *Out of Place: Englishness, Empire and the Locations of Identity*. Princeton, NJ.: Princeton University Press.
Baudrillard, Jean. 1990. *Fatal Strategies*. London: Semiotext(e)/Pluto.
Bonner, Frances. 2003. *Ordinary Television*. London: SAGE.
Brett, Judith, and Anthony Moran. 2006. *Ordinary People's Politics: Australians Talk about Life, Politics, and the Future of their Country*. North Melbourne: Pluto Press.
Broome, Richard. 2002. *Aboriginal Australians: Black Responses to White Dominance 1788–2001*. Sydney: Allen & Unwin.
Bryson, Bill. 2000. *Down Under*. Sydney: Random House.
Carter, David. 2006. *Dispossession, Dreams and Diversity: Issues in Australian Studies*. Frenchs Forest, NSW: Pearson Longman.
Caulfield, Annie. 1999. *The Winners' Enclosure*. London: Simon & Schuster.
Clarke, Robert. 2009. 'New Age Trippers': Aboriginality and Australian New Age Travel Books. *Studies in Travel Writing* 13 (1): 27–43.

Clarke, Robert. 2016. *Travel Writing from Black Australia: Utopia, Melancholia, and Aboriginality.* New York and London: Routledge.
Craik, Jennifer. 2001. Tourism, Culture and National Identity. In *Culture in Australia: Policies, Publics and Programs,* edited by D. Carter and T. Bennett. Cambridge: Cambridge University Press.
Dale, David. 2000. Please, Bill, No More Mr Nice Guy. *Sydney Morning Herald,* 8 July, 10.
Frow, John. 2007. UnAustralia: Strangeness and Value. Review of Reviewed Item. *Australian Humanities Review* 41 <http://www.lib.latrobe.edu.au/AHR/archive/Issue-February-2007/Frow.html> (accessed 15 October 2008).
Gans, Herbert. 1999. *Popular Culture and High Culture: An Analysis and Evaluation of Taste.* New York: Basic Books.
Gibson, Mark. 2001. Myths of Oz Cultural Studies: the Australian Beach and 'English' Ordinariness. *Continuum: Journal of Media and Cultural Studies* 15: 275-88.
Hage, Ghassan. 1998. *White Nation: Fantasies of White Supremacy in a Multicultural Society.* Sydney: Pluto Press.
Holland, Patrick, and Graham Huggan. 1998. *Tourists With Typewriters: Critical Reflections on Contemporary Travel Writing.* Ann Arbor: University of Michigan Press.
Human Rights and Equal Opportunity Commission. 1997. Bringing Them Home: Report of the National Inquiry into the Separation of Aboriginal and Torres Strait Islander Children from Their Families. Sydney: Human Rights and Equal Opportunity Commission (HREOC).
Inglis, K.S. 2001. *Sacred Places: War Memorials in the Australian Landscape.* Carlton South: Melbourne University Press. Original edition, 1998.
Kaplan, Alice, and Kristin Ross. 1987. Introduction. *French Yale Studies* 73:1–4.
Langton, Marcia. 1993. *'Well, I heard it on the radio and I saw it on the television …': An Essay for the Australian Film Commission on the Politics and Aesthetics of Filmmaking by and about Aboriginal People and Things.* Sydney: Australian Film Commission.
Lawson, Mark. 1993. *The Battle for Room Service: Journeys to All the Safe Places.* London: Picador.
Linquist, Sven. 2007. *Terra Nullius: A Journey through No One's Land.* London: Granta.
Lisle, Debbie. 2006. *The Global Politics of Contemporary Travel Writing.* Cambridge: Cambridge University Press.
Manne, Robert. 2001. In Denial: The Stolen Generations and the Right. *Quarterly Essay* 1: 1–113.
McCrum, Mark. 1997. *No Worries: A Journey Through Australia.* London: Sinclair-Stevenson.
McKee, Alan. 1997. 'The Aboriginal Version of Ken Done' - Banal Aboriginal Identities in Australia. *Cultural Studies* 11: 191–207.
Mickler, Steve. 1998. *The Myth of Privilege: Aboriginal Status, Media Visions, Public Ideas.* South Fremantle, WA.: Fremantle Arts Centre Press.
Morris, Meaghan. 1988. Tooth and Claw: Tales of Survival, and *Crocodile Dundee.* In *The Pirate's Fiancé: Feminisms, Reading, Postmodernism.* London: Verso.
———. 1993. At Henry Parkes Motel. In *Australian Cultural Studies,* edited by J. Frow and M. Morris. Sydney: Allen & Unwin.
———. 1998. *Too Soon, Too Late: History in Popular Culture.* Bloomington and Indianapolis: Indian University Press.
O'Regan, Tom. 1996. *Australian National Cinema.* London and New York: Routledge.
Rutherford, Jennifer. 1998. The Irish Conceit: Ireland and the New Australian Nationalism. In *Ireland and Australia, 1798–1998: Studies in Culture, Identity and Migration,* edited by P. Bull, F. Devlin-Glass and H. Doyle. Sydney: Crossing Press.
Scannell, Paddy. 1996. *Radio, Television and Modern Life.* Oxford: Blackwell.
Sinclair, Jennifer. 2004. Spirituality and the (Secular) Ordinary Australian Imaginary. *Continuum* 18 (2): 279–93.
Turner, Graeme. 1994. *Making it National: Nationalism and Australian Popular Culture.* Sydney: Allen & Unwin.

Chapter 11

"THE RIGHT SORT OF WOMAN"
British Women Travel Writers and Sports

Precious McKenzie Stearns

In 1850, Harriet Martineau recommended physically-demanding activities, such as swimming and rowing, for women. This position established exercise for women as a feminist cause. Many Victorians mistrusted supporters of the feminist movement and often looked upon them as unladylike or queer. During this period, Bessie Rayner Parkes lamented that "people endeavoured to check the physical power of their daughters as much as that of their minds" (Mitchell 1995: 106). By the 1860s, however, upper class and upper-middle class women actively participated in several sports. These women engaged in croquet, archery, yachting, fox-hunting, and riding. Such activities were appropriate or 'safe' for women, because they were "constrained by costume and custom" (Huggins 2004: 80). These leisure activities allowed both sexes the opportunity to play and watch together. Yet women were always expected to "behave like ladies" (2004: 80).

Critics viewed women's interest in sports as purely superficial. Ladies who shot drew great amounts of vocal criticism. Even Queen Victoria, in an 1882 letter to her daughter, Princess Victoria, expressed her distrust of athletic women. The Queen believed "it was acceptable for a woman to be a spectator, but only fast women shot" (Aitken 1987: 85). Lord Warwick, as late as 1917, remarked, "I have met ladies who shoot and I have come to the conclusion, being no longer young and a staunch Conservative, that I would prefer them not to" (1987: 85). Magazine cartoons frequently depicted women as "dizzy creatures with little interest in sport" (Huggins 2004: 81). Some thought women played sports only to flirt with men; the physiological need for demanding exercise was thought unnecessary for ladies. Contrary to many physicians' warnings and public scorn, some women of the late nineteenth century realized the importance of physical and mental exertion, and challenged the advice given by doctors, husbands, and even their own mothers. They sought adventure, excitement and usefulness in spite of public ridicule.

This article investigates two women travel writers who answered the feminist call initiated by Harriet Martineau, and who dared to transgress the boundaries of respectable Victorian leisure activities for women. The two women in this study, Lady Florence Douglas Dixie and Isabel Savory, were fortunate that their social status allowed them the freedom to travel and, thereby, participate in strenuous sporting activities. Women travel writers such as Dixie and Savory fueled the New Woman movement. Their readers often interpreted their achievements as travelers, adventurers, and sportswomen "as proof of female equality" (Stevenson 1982: 3). Thus, travel literature provided the impetus necessary for a Victorian sexual revolution. The New Woman of the late nineteenth century no longer contented herself with the home and hearth; she longed for the adventure and excitement that travel writing had inspired.

Women travel writers faced many of the same social constraints as their peers faced in England. For example, women were "not supposed to know or write about sex" (Mills 1991: 81). Many women travelers, however, journeyed with male companions or met with men during their expeditions. Yet few women travel writers mentioned any romantic rendezvous. Romance or sexual relationships seemed to have no place in their narratives. As Sara Mills states, "[W]ithin this stereotype, women are supposed to travel in order to paint butterflies and flowers" – certainly not hunt big game, scale mountains, navigate rivers, record data on foreign geographies, or have sex outside of marriage (1991: 81).

Nonetheless, some Victorian women travel writers did reject their submissive position as "good wives" at home in Britain, and many had no particular "affinity with domesticity" (Aitken 1987: 9). Indira Ghose adds: "what needs to be looked at in more depth is how notions of gender were bound up with hegemonic ideologies, and how women were both made an instrument of, and were complicitous with, the politics of imperialism" (Ghose 1998: 4). The woman question was truly more about women's place within the empire. How did their roles abroad differ from their roles in Britain? Oftentimes, as in Isabel Savory and Lady Florence Dixie's cases, travel allowed women to participate in activities that they did not usually engage in at home, such as sports and politics. When confronted with British imperial policies abroad, these women travelers often reflected on Britain's actions and the consequences of those actions. During the Victorian era, such outspokenness was not thought as fitting for true 'ladies.'

Revisiting the work of earlier feminist scholars, Janet Todd acknowledges "that gender as a social construct has real material effects on individuals and on a culture," and because "women experience the world differently from men ... politics, economics, and religion inform and influence the private and the domestic," as well as the public act of travel (Oliver 2000: 16). The woman traveler, such as Dixie, is contradictory because she, in certain instances, is subjugated, and "she has practiced the agency of constructing her subjectivity as well. So Woman is not merely a category, she is also a subjective positioning within which there is room for manoeuvre" (Smart 1992: 7-8). Joan Scott reinforces the importance of challenging a "reductionist approach that privileges gender to the exclusion of all other factors that construct identity" (Ghose 1998: 14). That being said, this article analyzes some of the social factors that contributed to the rise of women's participation in sports and adventure travel in the British Empire.

The travel writings of women differed significantly from those of male travel writers as male writers did not feel obligated to justify their choices of leisure activity. Victorian women travel writers often explained why they had broken from convention in order to travel and participate in sports. They recounted their actions in detail throughout their texts, so to retain a semblance of Victorian respectability. Deviations from the status quo, such as riding astride rather than sidesaddle, were justified by emphasizing concerns for their personal health and safety during their journeys.

For women in India, sport and travel differed significantly from travel to other parts of the British Empire. The power associated with guns and big game hunting set the British apart from the native Indian population. The British utilized the hunt as a reminder to the natives of British dominance and superiority. Although Anglo-Indian women were not allowed to hold positions in the imperial administration, women who hunted "secured for themselves a role in the empire and reordered ideas about the relationship between gender and imperialism" (Procida 2001: 488). Women who shot proved that they could defend themselves from ferocious Indian wildlife and also the supposed threat posed by hypersexual, immoral Indian men.

Aside from hunting and shooting, British women in India were active in many outdoor sports that their contemporaries in Britain were not. While in India, British women played polo and cricket. They were also training polo ponies and wearing men's clothing. Such practices did not signify "the revolt of the 'New Woman.'" Sport functioned as an essential marker of imperial femininity. British women in India managed to balance traditional traits and physical activities without erasing their womanhood. Activities such as riding and hunting were encouraged by their husbands (2001: 455). Sports for women flourished in India, because British women did not typically participate in philanthropy and democratic political activism (2001: 459-60). Nor were there many lending libraries or theaters for their entertainment (MacMillan 1988: 154). Outdoor sports, specifically hunting, became the vehicle in which British women participated in public life and empire-shaping.

It is important to note that British women were also challenging traditional conceptions of femininity while outside the British Empire. Women were traveling to China, Japan, Russia, and even parts of South America to escape the restrictive social codes of Victorian Britain. Lady Florence Douglas Dixie was one of these women. Lady Dixie, a daring horsewoman and close friend of the Prince of Wales, did not fit the mold of the traditional Victorian aristocrat; she was a sporting woman. She rode horses, hunted, and swam with men (Stevenson 1982: 45). Some thought of her as a moneyed rebel. After the birth of her son, Dixie decided to travel to escape the sameness of London society (Robinson 1990: 65).

Although Argentina was not controlled by the British, it was not a place completely removed from the British Empire, as it seems Dixie would have liked her readers to believe. By 1874, Britain provided a strong market for Argentine exports. Britain also invested in Argentina's transportation and communications projects (Shumway 1991: 279). British companies controlled three-quarters of Argentine railways by the 1870s (Davis and Huttenback 1988: 41). Moreover, after the invention of the refrigerated car, Argentine beef became a profitable export to Europe, as beef could remain fresh on the voyage across the Atlantic (1988: 70). Many Argentine administrators worked to make Argentina alluring for foreign investment and settlement. Specifically, President Nicolás Avellaneda "intensified the wars of displacement and extermination against the Indians, thereby making available large tracts of new lands for wealthy immigrants during the Conquest of the Desert (1870s-1884)" (Shumway 1991: 279). Dixie's travel book highlighted the natural beauty of the land and emphasized its vast untapped resources that might benefit the British Crown.

Dixie describes her journey across South America with her brother and her husband in her travel book, *Across Patagonia* (1881). As she crosses the frontier on horseback, Dixie ponders the slavery of women and the dangers that women face. The adventurers encounter prairie fires, food shortages, and thrilling hunts. When Dixie participates in the hunts, she questions English tradition. The chase is exciting. She loves the open air. Yet the slow, drawn-out deaths of the animals disgust her. Her time spent in the open South American air does not masculinize her; rather, it provides a lesson in self-reliance that transcends gender boundaries. Her womanhood, she believes, is strengthened because she has learned to provide for herself and survive in difficult circumstances. Her femininity is emphasized because she develops sympathy and compassion for the animals that she hunts.

Upon her return from big game hunting in South America, Dixie becomes an outspoken critic of blood sports and a champion of animal as well as women's rights. Dixie redefines womanhood. She proves that she is brave and self-reliant, and yet she is nurturing and compassionate, as a British woman should be.

Isabel Savory's travel book, *A Sportswoman in India: Personal Adventures and Experiences of Travel in Known and Unknown India* (1900), differs significantly from Dixie's text. Dixie travels to escape rigid British gender roles, and, for the most part, she succeeds in Patagonia. Savory travels by boat, train, elephant, and horse in order to experience the "greatest jewel in the British crown" and to witness British imperial accomplishments. Savory claims that she travels to expand her mind and develop herself. To broaden her education, she catalogues India's riches, explores causes of poverty, and ponders international relations. Her opinions are diverse and fluctuating. She criticizes British colonial policy regarding Kashmir, even as she denigrates Indian social customs. As a British woman in India, Savory seems torn between defending her homeland and criticizing its policies in India.

Isabel Savory's narrative differs from earlier travel writers in that she voices her concerns surrounding women's education and health without the common rejoinder, 'but I am only a woman.' Significantly, Savory's hunting narrative serves "as a recruiting and advice manual for prospective female big-game hunters" (Gates 1998: 202). Isabel Savory does not attempt to 'whitewash' or 'prettify' the genre of the hunting narrative. She recounts her experiences "as examples to other venturesome women, the implication being that women can emerge as the heroines of their own adventure narratives" (2002: 308-9). By crafting herself as a venturesome woman, Savory re-envisions the traditional British masculine adventure tales that were popular during the nineteenth century. Isabel Savory, as the bold big game hunter, appears as competent as the heroes from the stories of Rider Haggard, Robert Ballantyne, G.A. Henty, and Rudyard Kipling. *A Sportswoman in India* asserts women's need for action and adventure. Savory hopes to prove that travel, sport, and comradeship is a "perfectly legitimate aspiration for the female," as it is for men (McKenzie 2005: 549).

Perhaps Savory's self-confidence marks the difference between her and earlier women travel writers. Writing in 1900, Savory enjoys the fruits of earlier struggles for women's rights. Isabel Savory's travel narrative transcends the scope of traditional nineteenth-century accounts and, instead, offers innovative observations of travel and life in India over the course of one year. Like Dixie, Savory finds travel beneficial to women. She writes, "[T]ravel has many advantages, of course; nothing appeals to mankind like 'change,' or better satisfies the restlessness felt at some time or another by every human being" (Savory 1900: 403). Traveling offers women an opportunity to escape what Florence Dixie labels as the *ennui* felt at home. The act of travel opened the traveler to new experiences, and with this came increased self-knowledge and self-confidence. Savory expands on this theory: "[E]xperience means a variety of things: it includes the development of the perceptive powers, dependence upon self, and a wider knowledge of self; it inculcates generous views; it causes, in short, a great mental expansion" (1900: 404).

The late nineteenth century's enthusiasm for health and fitness contributed to the women's movement in that some women left the domestic sphere in hopes of improving their mental and physical health. The act of travel and the challenges that arose while out in the empire led to reevaluations of ideological constructions of womanhood. The writings of Dixie and Savory prove that as women tested their physical limits in adventure travel and sports, they experienced increased self-confidence in their physical and mental capabilities. In turn, British women readers began to question and reject the patriarchal ideal of the Victorian angel in the house.

Sports, specifically hunting, liberated British women from domesticity. Not only did women's participation in the hunt renegotiate gender roles, but hunting reinforced alliances between the British and Indian aristocracy (Procida 2001: 476). Big game hunting was justified by British imperialists because they thought, "Indians would have been at the mercy of rampaging elephants and voracious tigers without the beneficent protection of well-armed male and female imperialists" (2001: 477). The hunt then functioned as yet another example of the British assuming the *white man's burden*.

While most British women travel writers did not directly address sex, perhaps the hunt served as a substitute for sex. John M. MacKenzie in *The Empire of Nature: Hunting, Conservation and British Imperialism* (1988) interprets the hunt as "a pseudo-sexual act" because

> hunting works are full of descriptions of the physical agonies of the Hunt, of "the exaltation no civilised world can supply," the tensions induced by great risk, and the ecstasy of release when the hunter prevails and stands over his kill. Some of this can readily be interpreted as sexual sublimation, while the sexual analogue of the gradual building up of the chase and the orgasmic character of the kill had long been recognised in writings about and pictorial representations of the Hunt. (MacKenzie 1988: 42-43)

Hunting narratives written by women could very well be read as powerful releases of emotion and exertion metaphorically linked to having sex.

To build courage, vigor, and sexual potency in the next generation of British men, "[b]oys should be given 'manly' rather than 'refined' books, which should help to create 'manly boys, uttering manly thoughts, and performing manly actions'" (1988: 46). This practice contributed to the development and proliferation of ideologies of Victorian masculinity. Killing wildlife became an integral component of moral instruction (2005: 548). However, "killing was far from being the whole of it" as "[f]ield sports prepared the boy for his life's journey which would require stoicism, perseverance and robustness" (2005: 547). These traits were certainly thought appropriate for men, but not for women.

Well into the 1890s, shooting in Britain remained primarily a masculine pursuit. This situation was largely due to the "few practical, technical or inspirational books for girls" (2005: 551). Because of this limitation, "women shooters were disadvantaged from childhood by custom, literature, training and opportunity" (2005: 551). Moreover, if the hunt was "pseudo-sexual," to allow women to participate would be to expose and even encourage raw, unharnessed female urges and desires. Savory challenged the masculine hegemony surrounding the hunt. She wrote "in an effort to encourage lady shots." From Savory's perspective, male hunting narratives were "excessive and repetitive." She, on the other hand, "killed sparingly but effectively – a 'selective' rather than 'non-selective' killer!" (2005: 549)

Jane Robinson, in *Wayward Women: A Guide to Women Travellers*, argues that Isabel Savory "revels in a polite orgy of fox-hunting, pig-sticking, and tiger shooting, with a touch of Himalayan couloir-climbing and conventional sightseeing thrown in" (Robinson 1990: 74). To read Savory's text as a blood orgy detracts from its more subtle social message. Savory, herself rejecting British domestic confinement after marriage, views hunting as an act of liberation for women. By 1900, women hunters were "not quite beyond the pale" of respectability, since "most of them were rich" (1990: 62-3). Like Dixie, Savory's wealth, to some degree, insulates her from social criticism, thus, safeguarding her respectability. Nevertheless, she challenges conventional patriarchal ideologies surrounding women's health and women and bloodshed. Florence Dixie and Isabel Savory demonstrate to readers that women could enjoy and excel at sports – even blood sports (although Dixie would later change her position on blood sports). Sport and adventure travel proved a fascinating lifestyle that many young British female readers hoped to emulate.

Pig-sticking in India provided an opportunity for the British to demonstrate their culture of courage. Such a demonstration of bravery illustrated to Indians that British power would protect them from both human and animal predators. Similar to a character that one might find in a Haggard or Kipling story, Savory readily assumes the imperial *white man's burden* by ridding India of ferocious beasts. She transforms herself into the archetypal imperial adventurer straight out of fiction, when she describes her journey through Lahore on the back of an elephant under the glaring Eastern sun. Later, she dines near water under Indian moonlight (Savory 1900: 34).

The Right Sort of Woman

By 1900, British power was not stable, and Savory realized it. Savory's hunting narrative served to reinforce notions of British determination, courage, and dominance, but, paradoxically, "hunting represented an increasing concern with the external appearance of authority, the fascination with the outward symbols serving to conceal inner weakness" (MacKenzie 1988: 171). Modernist writers such as Joseph Conrad and E.M. Forster would revisit the weaknesses of imperial Britain and question British attempts to assimilate native populations into the empire.

During Savory's time in India, she abandons civilized British conventions regarding gender, since she spends a considerable amount of time out of doors. For refined British women, doing so could appear rough or coarse. Yet for Savory and Dixie, an active, outdoor lifestyle ensures good health and contentment. Savory remarks that life in England differs significantly from her life in India:

> [A]t home we do not take enough advantage of fine weather: away in the Himalayas we lived out of doors. Riding out the first thing in the morning two or three miles, to breakfast on some wild hill shaded with great deodars and carpeted with fern, we would find a quiet corner in the winding path, moss to lie upon, steep rock rising sharply behind us, and in front of us vistas of tree-tops and undergrowth half hidden down the precipitous hillside. Across the valley below we looked on to the mountains topped with snow, dazzling in the early sunlight. (Savory 1900: 82-3)

As it did for Dixie, nature offered Savory time away from rigid social expectations. Nature did not destroy their femininity; rather, contact with nature invigorated and refreshed their spirits, thereby transforming them into resilient, healthy, capable women – New Women.

Feelings of well-being were some of the most important benefits gained from sports and travel. With the realization that many women were reluctant to participate in outdoor sports, Savory asserts, "[w]omen do not shoot with their husbands and brothers nearly as much as they might do, provided they are the right sort of women. Of course, there are *women* and *women*: but in the present day, when so many of them care for a free life, I wonder that the majority of those should still live a conventional one" (1900: 130-1). Savory perceives a disconnect between the rhetoric and practices of turn-of-the-century women. Contrary to the Victorian ideal of the angel in the house, Savory asserts that the 'right sort of woman' in the British Empire is active, strong – and upper-class white. Because of the joy that she experiences in the hunt, Savory cannot fathom why women would hesitate to participate in sports, or, why they would choose to remain confined in the home.

Precious McKenzie Stearns

The dangerous life of a sportswoman seems the only life suitable for Savory. Personal pleasure and a sense of accomplishment in the field are her rewards. Nothing makes her happier than the thrill of the chase. Her narrative demonstrates that women could enjoy and be successful in the hunt. Moreover, guns allowed women to be "the true partners" of men, since skill with firearms was more important that one's sex or brute strength (Procida 2001: 470). On one occasion, Savory scrambles up and down mountains with her female friend, M., in pursuit of a goat. As they track the animal, the ladies discuss tea and mutton sandwiches. M. successfully kills one goat, leaving the ladies "thoroughly pleased" with themselves (Savory 1900: 434-6). The adventure created by hiking, tracking, and killing rescue the two women from the traditional domestic sphere. Rather than order cooks around the kitchen, Savory and M. exercise their bodies and develop personal agency. The women successfully capture their supper.

According to Savory, to be a true sportswoman requires six basic principles. She defines a real sportswoman as someone who has an appreciation of the "free camp life," is a "fair shot," never participates in unsportsmanlike action, prefers "quality to quantity in a bag," and is a "keen observer of all animals, and a real lover of nature" (1900: 140). Such a characterization of the true sportswoman encourages British readers and hunters to see themselves as heroes. The cult of the heroic sportsman (and sportswoman) was an avatar of the cult of British military hero worship. In *Britons: Forging the Nation, 1707-1837*, Linda Colley asserts that British hero worship stemmed from Britain's history of political crisis. After the French Revolution the British elite "took more than usual care to appear scrupously religious and morally impeccable" (Colley 2005: 192). Britons identified "themselves with military achievement on the battlefield and with a cult of civilian heroism at home" (2005: 192). The cult of heroism was also carried to the sporting field, as British men worked together to battle nature. Savory's definition of a sportswoman, as well as her narrative chock full of substantial examples, asserts that women, like men, are capable of carrying on the British tradition of heroism. The result is heroism and outdoor sports metastasized into a core component of British national identity.

More so than South America, British India became the theater for British heroism. British India during the late nineteenth and early twentieth centuries witnessed a "partnership between men and women as imperialists in the masculine mold, rather than an antagonistic polarity between adventuresome men and ultra domestic women" (Procida 2001: 462). Ultra-feminine women proved themselves unsuitable to the harsh living conditions in India. To survive in India, women redefined their ideas of femininity. The image of the *mem-sahib* – a beautiful, wealthy British woman, usually found lounging on a sofa in the stifling Indian heat – was a stereotype surrounding British femininity in India. Savory corrects such a stereotype. Contrary to the popular beliefs about the beautiful, weary *mem-sahib*, Savory claims such creatures do not exist, nor do the "dyspepsia–and-liver individual." She describes the *mem-sahib* as an "energetic, tennis, Badminton, calling and riding – sometimes sporting – creation" (Savory 1900: 338). Alterations in fashion, along with modifications to the saddle, led the way for social reform for women (Procida 2001: 462). British women in India were perhaps the most liberated in the empire, because they were allowed and even expected to participate in sports. Indian servants tended to household matters, while the *mem-sahibs* sported. British angels were out of the house.

By the late nineteenth century, privileged British women traveled to the far reaches of the empire to escape the *ennui* that they experienced at home. Women traveled across the world "with the object of seeing other sides of that interesting individual, man, other corners of the world, other occupations, and other sports" (Savory 1900: 201-2). Women proved that they could manage "a little discomfort for the sake of experience" (1900: 201-2). The experience and knowledge gleaned from adventure travel, Savory finds, "is a complete change for an ordinary English girl; and it is very easy to find every scope for developing self-control and energy in many a 'tight corner' if such occasions are sought for" (1900: 201-2). The challenges found in travel allowed British women to demonstrate that they were resilient, strong, and resourceful. Such sentiments fueled the women's movement, as they allowed young women readers to witness active women faced with dangerous situations. By the early twentieth century, increasing numbers of women were participating in organized sports, shooting for pleasure, and demanding equal rights with men. The women's movement sought the inclusion and active participation of women in the public sphere. By the turn of the century, women were clearly 'craving to do something.'

Travel and sports provided immeasurable benefits for women, as Savory and Dixie realize. Savory states that her purpose in *A Sportswoman in India* is to demonstrate the benefits of both travel and sports. Therefore, the central tenet of her book is:

> [t]o see more is to feel more; and to feel more is to think more. Travel teaches us to see over our boundary fences, to think less intolerantly, less contemptuously of each other. It teaches us to overlook the limitations of religions and morality, and to recognize that they are relative terms, fluctuating quantities, husks round the kernel of truth. Travel dismisses the notion that we are each of us the biggest dog in the kennel. (1900: 404)

Savory thought that people should travel more, so they would not feel alienated. Most importantly, she emphasized that knowledge of other cultures bred tolerance (1900: 405). Unfortunately, only a small percentage of the English population had the economic means to travel overseas (1900: 408). The privileged few who actually experienced foreign travel oftentimes followed in Savory's footsteps: they recorded their experiences in travel books and magazines. Women situated within the British elite were not only traveling overseas, but they were also "hunting with dogs, shooting elephants and rhino and displaying 'true nerve'" (McKenzie 2005: 550).

Dixie's *Across Patagonia* and Savory's *A Sportswoman in India* focus on the benefits of an active life. Because of the positive accounts of athleticism by women travelers, female health became a prominent topic in society. By the early twentieth century, women had increased their access to physical pursuits. Girls' schools taught physical education – in proper dress, of course (Gleadle 2001: 183). Popular magazines, such as *The Girl's Own Paper*, featured athletic activities for young women (Mitchell 1995: 105). Medical doctors realized that "exercise improved women's health and childbearing capacity" (1995: 106), yet this realization competed with increased anxiety over the loss of feminine charms, due to the perceived masculinization caused by exercise and sport. Nevertheless, by the early twentieth century, British women seized opportunities to participate in athletics, thereby redefining themselves in the debate over women's rights. Indeed, sports and travel inspired women to escape the traditional domestic sphere in order to actively participate in empire-building.

By World War II, however, "hunting and shooting declined as imperial power waned" (MacKenzie 1988: 195). Many writers during this time noted that recruits to India were no longer participating in traditional shooting parties. John M. MacKenzie argues that modern firearms affected British sentiments regarding the hunt (1988: 195-6). Administrative confusion and civil unrest during the interwar years made it difficult for British officers to find time for large-scale hunting expeditions (1988: 196). Despite the decline of hunting in India, some women received professional firearms training, so they might better defend themselves during times of civil unrest (Procida 2001: 486).

British women's use of guns after World War I indicated that they were ready to defend the empire against perceived threats, even if it meant sacrificing traditional notions of femininity (2001: 487). Women who were ready to defend themselves and the empire were admired. An early twentieth-century observer of British India remarked that the "right sort" of woman for India was the "Good-All-Around-Woman." This woman would "exhibit a courage and spirit, a presence of mind, a calmness in danger or difficulty, *which any man might be proud of*" (2001: 464). India and South America were not places for "the poor creature ... who faints at the sight of a cut finger, shrieks if a mouse runs by; who becomes 'muddled' at a crisis, who whines and whimpers when anything dreadful happens" (2001: 464).

Adventure travel writing became the arena in which Britons reevaluated traditional expectations of women. Savory and Dixie's travel narratives contributed to this reevaluation of womanhood. By the interwar years, "the empire was thus a place for good sports, chums, and comrades-in-arms of either sex" (2001: 464). Sports played a large role in transforming the patriarchal social order of the British Empire. Sportswomen, such as Isabel Savory and Lady Florence Dixie, increased women's access to and participation in the public sphere. They reshaped traditional ideas of the weak, submissive Victorian woman and, instead, recreated Victorian womanhood as bold, confident, and ready for action.

Acknowledgment

The author would like to thank Kathleen Paul of the University of South Florida for her comments on earlier versions of this manuscript.

Precious McKenzie is an associate professor of English at Rocky Mountain College. She teaches Travel Literature, British Literature, and Environmental Literature as well as writing courses. She is the author of poems, short stories, and academic articles as well as books for children. Her research interests include gender and imperialism and transatlantic studies. She is a member of the Rocky Mountain Modern Language Association, AWP, and NCTE.

References

Aitken, Maria. 1987. *Women Adventurers: Travelers, Explorers, and Seekers.* New York: Dorset Press.
Colley, Linda. 2005. *Britons: Forging the Nation, 1707-1837.* New Haven: Yale University Press.
Davis, Lance E. and Robert A. Huttenback. 1988. *Mammon and the Pursuit of Empire: The Economics of British Imperialism.* Cambridge: Cambridge University Press.
Dixie, Florence Douglas. 1881. *Across Patagonia.* New York: Worthington Co.
Gates, Barbara T. 1998. *Kindred Nature: Victorian and Edwardian Women Embrace the Living World.* Chicago: University of Chicago Press.
Gates, Barbara T., ed. 2002. *In Nature's Name: An Anthology of Women's Writing and Illustrations, 1780-1930.* Chicago: The University of Chicago Press.
Ghose, Indira. 1998. *Women Travellers in Colonial India: The Power of the Female Gaze.* Oxford: Oxford University Press.
Gleadle, Kathryn. 2001. *British Women in the Nineteenth Century.* New York: Palgrave.

Hervey, H.J.A. 1913. *The European in India.* London: S. Paul & Co.
Huggins, Mike. 2004. *The Victorians and Sport.* London and New York: Hambledon and London.
MacKenzie, John M. 1988. *The Empire of Nature: Hunting, Conservation and British Imperialism.* Manchester and New York: Manchester University Press.
MacMillan, Margaret. 1988. *Women of the Raj.* London: Thames and Hudson.
McKenzie, Callum. 2005. "'Sadly Neglected'- Hunting and Gendered Identities: A Study in Gender Construction." *The International Journal of the History of Sport* 22, no. 4: 545-562. doi: 10.1080/09523360500122848
Mitchell, Sally. 1995. *The New Girl: Girls' Culture in England, 1880-1915.* New York: Columbia University Press.
Mills, Sara. 1991. *Discourses of Difference: An Analysis of Women's Travel Writing and Colonialism.* London: Routledge.
Oliver, Kathleen M. 2000. "(En)Gendering Silence: Women and Silence in the English Novel, From Richardson to Austen." PhD diss., University of South Florida.
Procida, Mary A. 2001. "Good Sports and Right Sorts: Guns, Gender, and Imperialism in British India." *The Journal of British Studies* 40, no. 4: 454-488.
Robinson, Jane. 1990. *Wayward Women: A Guide to Women Travellers.* New York: Oxford University Press.
Savory, Isabel. 1900. *A Sportswoman in India; Personal Adventures and Experiences of Travel in Known and Unknown India.* London: Hutchinson.
Shumway, Nicolas. 1991. *The Invention of Argentina.* Berkeley: University of California Press.
Smart, Carol. 1992. "Disruptive Bodies and Unruly Sex: The Regulation of Reproduction and Sexuality in the Nineteenth Century." Pp. 7-32 in *Regulating Womanhood: Historical Essays on Marriage, Motherhood and Sexuality*, ed. Carol Smart. London: Routledge.
Stevenson, Catherine Barnes. 1982. *Victorian Women Travel Writers in Africa.* Boston: Twayne Publishers.

Conclusion
Interminable Journeys

Pramod K. Nayar

Allow me to set, as the point of departure, an autobiographical starting block. My own experience of travelling through *Journeys* has been one of constant wonder: at the range of geo-cultural locations examined, the retrieval of texts, and the inherently interdisciplinary nature of the work on the genre. Most rewarding has been the consistent rigor of the methodology in the essays over the years, the discovery—a prototype travel trope, if there ever was one!—of newer texts, images, and ideologies of mobility, from the colonial to the postcolonial, medieval to the contemporary, that *Journeys* has offered to the reader over the last two decades. Aesthetics, narrative modes, images and drawings, maps and topographies, politics, and poetry have all figured, at some point or the other, in the pages of the journal.

This collection, as befits a volume on travel writing, traverses multiple spaces, geographical regions, temporal frames, and cultural perceptions. The essays here deal with imaginative geographies, discourses of travel, the function of stereotypes, cultural productions, and knowledge systems. It "circles the globe", as the editors—Maria Pia Di Bella and Brian Yothers—put it in the introduction.

Travel alters the shape of the world by altering the perceptions of the world. Colonial travel, discovery trips, military expeditions, and tourism informed the people back home of the outside world when the travellers returned and circulated their narratives. Images, material objects, stories, and people disseminate as "travelling matter." Genres as diverse as travel guides and photographic memoirs, from the travel diary to the memoir, are part and parcel of the travel apparatus, as we think of the practices and discourses that constitute travel.

Travel/travail: as an etymologically linked binary, the commonplace terms for "journeys" automatically inclines "travel" towards "suffering." Recreating, via travel, a trauma from the past is travail of the footsteps variety. When, for instance, Saidiya Hartman in *Lose Your Mother*, seeks to recreate the slave trade route across Africa in an instantiation of Peter Hulme's "footsteps travel," one recognizes the interminable nature of historical travails. But it is not just trauma and travel that come together in the past, present, and future of travel writing, or in academic work on the same subject.

With more and more explorations of earlier journeys—the recreation of the Silk Route comes immediately to mind—being undertaken, footsteps travel is a major genre worth exploring. Footsteps travel is about memory, of course, and the politics of memory is inseparable from mechanisms and strategies devised to inform and influence a certain kind of travel through memory lanes. When Maria Pia Di Bella examines the Berlin street memorials devoted to the preservation of the Holocaust memories, she proposes that "today's generation looks at the history of its society as a heritage whose elements receive their moral and political significance through the judgment of the generation that follows—their own." Memory travels from one generation to the next, is judged from one generation to the next, with certain kinds of "carriers" of both memory and trauma With each stage of the journey, as Hartman shows and as the essays in this section document, the import—semantic, symbolic and affective—of the journey and the travellers from *that* past, changed. For African Americans, Germans, descendants of indentured labour who travelled under duress, and other such inheritors of bruised memories, embodied in exhibitions, fiction, travel accounts and recreated journeys, the "acts of memory," as I have argued in the case of Hartman, "are acts that seek a 'memory citizenship' in problematic and complicated ways." Hence, footsteps travel, museum cultures, mnemonic practices and discourses that examine colonial travel, slave travels and the Middle Passage, the Underground Railroad and such, are potentially a great resource not only to study travel writing but also memory cultures.

The phrase "gathering material for future nostalgias," as Vikram Seth describes his travel in his memoir, *From Heaven Lake*, captures the link between travel and memory. Seth's emphasis is not on the consumption of memory in a museum or through footsteps travel but about the very self-conscious *production* of memory, for future use, *as one travels*. The diary, the journal, and now the selfie are devices of memory-making. Modes of memory-making alongside the experience of travel, and how the two inform each other, especially in the age of the selfie and Instagram, is worth examining.

Apparatuses that manufacture and regulate mobility regimes are a key resource for the study of travel writing. Academic work has examined travel guides and books as instances. But there is much more. For instance, the role of the Thomas Cook agency or the massive P&O liners in the 19th century in determining and shaping the colonial travel experience would be worth studying. Advertisements and brochures announcing exotic locations and the ben-

Conclusion

efits of travel, memorabilia, and souvenirs are as integral to the travel experience as the physical journey, and merit attention too.

People travel, but so do other forms of life and materials. Take for instance, the story of one such animal in Glynis Ridley's *Clara's Grand Tour: Travels with a Rhinoceros in Eighteenth-Century Europe*. How did this rhinoceros travel halfway across the world, from Assam (India) to Europe? Ridley records the reception and the elaborate apparatuses that made the journey possible. Exotic animals arriving in Europe brought the world into their ken, so to speak. Records show the import of various exotic animals into the European spectacle and entertainment industry from the 18th century. The transportation of animals of various kinds has not been studied—perhaps because the animals did not keep records ?—but ought to be. What rhetoric of travel framed the animals and their visual consumption by the armchair travellers of London? As people like Lucille Brockway, David Arnold, and Ann Colley have shown, specimen and museum collections—Ashmolean, the British Museum, the Leiden and Heidelberg museums, the Kew Gardens—in the European countries were possible through the travel of intrepid collectors. But objects—living, non-living—also travelled. The circulation of material cultures indexed the colonial, mercantile and scientific projects of the imperial age. This material "shrinking"—if that is what it is—of the world into the emporium, the museum and the private collection is a project worth thinking about as a form of travel too.

A certain kind of travel-culture popularized as adventure tourism and survival games has emerged in the recent past. In earlier eras, mountaineering and exploration of hitherto "undiscovered" (the term is contingent on a binary, of the European as discoverer and the non-European as the discovered) spaces in the imperial age were instantiations of risk-taking and culture of discovery. These were informed, argues Elaine Freedgood, by discourses and attitudes of Victorian "cultural masochism," where fear, pain and suffering (experienced in such travel) evoked *voluntarily* makes and remakes the world. The big-game hunter and the explorer of dangerous places have their *then* conditions replicated today in the form of survival games and adventure tourism, where competitions and extreme sporting events set in inhospitable and harsh terrains are exercises in a new form of world-making. This new form of travel, commercialized, monitored, and technologically supported as never before, is a domain open to scrutiny.

The personal experience of travel/travail is foregrounded through all travel writing. This truism has, however, a potentially explosive—I use the term advisedly—aspect to it. Forms of travel and transportation that triggered civil society movements, large-scale campaigns, and even nation-wide events may not be within the purview of what we understand as travel writing, but, as noted above, they are instantiations of apparatuses of travel and determine mobility regimes. Think of: Mahatma Gandhi thrown off a train in South Africa, B.R. Ambedkar experiencing caste-based discrimination and abuse on a train as a child, Rosa Parks refusing to vacate a seat on a bus. Having experienced trauma proceeding from a structural condition of inequality, on a form of transport, each of these triggered socially transformative movements that shaped the modern world. The travels of Gandhi across India—by train, a British invention!—to unite the country against the British, Martin Luther King's many journeys (and the famous bus boycott), Ambedkar's travels through India to campaign for the oppressed classes are, in many ways, interminable journeys. The journeys they then undertook changed their respective societies forever, the effects of which are still felt in the form of civil rights campaigns and the unending search for justice. There may not be travel writing here, but there is a mobility regime in action, and that is worth studying.

How do concepts travel? Critics as diverse as James Clifford and Edward Said have spoken of "travelling theory." How does theory travel? How do key concepts from say, Greek philosophy travel via French theory, into English literary criticism? How does translation work with, or engender, the travelling of concepts and ideas? The navigation of a word or idea—let us take the most common ones, say, "ideology" or "trauma"—traverse the critical terrain from the strangest of provenances to cosmopolitanized, global acceptability, and respectability? How does a concept intersect, inflect, and inform home-grown ideas and concepts in, say, African philosophy or literatures in Indian language? How do we map the journey of a word in this cosmopolitan world of intellectual work?

It seems appropriate, given the circumstances, to end with a different version of travel: the digital. "Surfing" now has more than one meaning, but retains the semantic foundation of implying movement. Are travels in cyberspace, *travel*? As we leave digital footprints—the subject of numerous thrillers and pop fiction involving cyberspace tracking and stalking—through the digital space when we click and click, is there an experience of travel? What does the haptics of the click or the visual appeal of cameras zooming in *do* to our experience of travel in cyberspace? With images of the information highway and the troping of traces as footprints, the metaphors of travel have clearly spilled over into the digital world. How does the cyber-walker walk? How does the surfer travel? The experiential shift involved is surely something to engage with.

I wrote this conclusion during the COVID-19 pandemic in 2020, when all forms of journey, short/long, proximate/distant, adventurous/essential, have been curtailed or even terminated, with fears of circulating pathogens (in the double sense of pathogens that circulate but also travel that enables us, as vectors, to cause the circulation of pathogens). Amidst this, the horrific forced on-foot journeys of migrant labor across thousands of kilometers to their hometowns as their employment (and income, and food, and shelter) suddenly came to an end with the virus-driven lockdown, also brought to attention the differential mobility regimes still in place. The conclusion did, given the "terminating" connotations of the term "conclusion," resonate with the apocalyptic and doomsday rhetoric around the pathogenic year 2020. Travel writing, a solution to the absence of any other form of travel (other than surfing the www), seems more resplendent in the impossibility of its embodied enactment today. Let us hope the journey of the pathogen ends quickly, and travel can resume. As it should, for the world's travels are inevitable and interminable.

Pramod K. Nayar teaches at the Dept of English, University of Hyderabad, India. His most recent books include, *Indian Travel Writing in the Age of Empire, 1830-1940* (Bloomsbury 2020), *Ecoprecarity: Vulnerable Lives in Literature and Culture* (Routledge 2019), *The Extreme in Contemporary Culture* (Rowman and Littlefield, 2017) and *Bhopal's Ecological Gothic: Disaster, Precarity and the Biopolitical Uncanny* (Lexington 2017), besides essays in *Narrative, a/b: auto/biography studies, Celebrity Studies*. Forthcoming is a book on the Human Rights Graphic Novel from Routledge.

References

Freedgood, Elaine. 2000. *The Victorian Writing of Risk: Imagining a Safe England in a Dangerous World*. Cambridge: Cambridge University Press.
Hartman, Saidiya. 2011. *Lose Your Mother: A Journey Along the Atlantic Slave Route*. New Delhi: Navayana.
Hulme, Peter. 1997. "In the Wake of Columbus: Frederick Ober's Ambulant Gloss," *Literature and History* 6 .1: 18–36.
Nayar, Pramod K. 2013. "Mobility Migrant Mnemonics and Memory Citizenship: Saidiya Hartman's *Lose Your Mother*." *Nordic Journal of English Studies,* 12.2: 81-101.

Index

A

Abolitionism, 33
Aboriginality, vi, 7, 168-170, 173, 175, 177-184, 186-187
Afghanistan, v, 1, 5, 70-74, 76-84
Agamben, Giorgio, 135
Aleppo, vi, 1, 7, 150-167
Anburey, Thomas, 64, 68
Anthropology, 1, 3, 128-129
Argentina, 1, 6-7, 191, 200
Arnold, Matthew, 28, 137-138, 149, 203
Asia, 1-3, 6, 35, 40, 43, 49, 74, 76, 107, 154, 166-167
Assam, 203
Auschwitz, 7, 9, 20
Australia, vi, 1, 7, 168-187

B

Balkanism, 5, 86, 101
Balkans, 1, 5, 85-88, 96-97, 99-100, 102, 104-107
Ballantyne, Robert, 192
Bataille, Georges, 134
Baudelaire, Charles, 138-139, 148
Baugh, Bruce, 134-135, 149
Bebelplatz, vii, 14-15, 24-25
Benin, 28, 30-31, 34
Berlin, v, vii, 1, 3, 8, 13-20, 22-27, 88, 106, 202
Berlin Senate, 15, 18, 22
Berlin Wall, 13, 18, 25, 88
Blanchot, Maurice, 134-135, 149
Blood sports, 191, 194
Boltanski, Christian, 14, 18, 25
Bonn, 24, 27

British Empire, 75, 190-191, 195, 199
Bryson, Bill, 168-175, 177-180, 184-186
Buchan, Alexander, 57
Bulgaria, 85, 92-93
Bush, George W., 73, 80, 84, 119
Buzard, James, 2, 9, 136, 149
Byron, Robert, v, 5, 56-57, 70-71, 74-76, 81-82, 107

C

Caley-Webster, Herbert, 112, 127
Cannibalism, 108-110, 113-114, 124, 127, 129
Cape of Good Hope, 56, 69, 151
Catholicism, 152, 156, 159, 165, 184,
Caulfield, Annie, 168-173, 175-178, 180-181, 184, 186
China, v, 4, 35-37, 39-40, 42, 45-46, 48-50, 61, 68, 126, 191
Christianity, 38, 83, 152, 167
Colley, Linda, 166, 196, 199, 203
Colonialism, 2, 7, 42, 75, 80-81, 83, 107, 128-129, 168, 181, 185, 200
Confessions of a Young Man, 133, 149
Cook, James vii, 53, 57-59, 65-66, 68-69, 202
Cosmopolitanism, 134, 149, 155
Crane, Nicholas, 88-90, 92, 94, 106
Croatia, 97

D

Dalrymple, Alexander, 57
Dalrymple, William, 2, 57-58, 69
de Man, Paul, 141, 149
Demnig, Gunter, vii, 14, 19-22, 27

Index

Description of a Slave Ship, vii, 30-34
Désoeuvrement, 134-135
diaspora, v, 4, 28-29, 31, 33, 82
Dickens, Charles, 135, 140, 149
difference, v, 4, 18, 53, 55-57, 59-61, 63-69, 86-90, 92, 94, 96-100, 103-105, 107, 110, 134, 152, 169-170, 177, 192, 200
Digital travels, 204
Dimic, Moma, 97, 101-104, 106
Dixie, Florence Douglas, 189-195, 197-199
Drake, Francis, vii, 61-63, 67-69

E

East India Company, 66, 151
Egypt, 57, 69, 154, 166
Eliot, T.S. (Thomas Sterns), 148
Ellis, Henry Havelock, 128, 142
Empty Library, vii, 14-15, 24-25
England, 4, 137-138, 140, 142, 151-155, 160-161, 164-165, 176-177, 189, 195, 200, 205
Europe, vi, 1-3, 5-6, 13, 24, 26-28, 35, 53, 74-75, 85-89, 91, 93, 95-97, 99, 101, 103-107, 137-138, 151-152, 154, 156, 159-161, 163, 165-166, 191, 203

F

Fanon, Frantz, 98, 106
Fanon, 98, 106
Fiji, 108, 118-119, 122, 127
Footsteps travel, 202
France, 139, 141, 154, 160, 164, 168
Fussell, Paul, 2, 9, 74-75, 83

G

Gautier, Théophile, 138
Genocide, 4, 8, 14-16, 24, 26, 83
Germany, 14, 17, 19, 24-25, 28, 61, 69, 91, 101, 126, 154, 164, 166, 184
Ghose, Indira, 189-190, 199
Gilpin, William, 57-58, 69
Girl's Own Paper, The, 198
Gregory, Derek, 71-74, 80, 83
Greenblatt, Stephen, 2, 9

Griffiths, Julius, 155, 161-162, 166
Grimshaw, Beatrice, 110-111, 128
Grunewald Platform 17 (Gleis 17 Mahnmal), 17, 26

H

Haggard, Rider, 94, 138, 192, 194
Hartman, Saidiya, 201-202, 205
Hazoumè, Romuald, 28, 30-31, 33-34
Heine, Heinrich, 14
Henley, William Ernest, 145
Henty, G.A., 192
Hero worship, 196
Heymen, John, 153-154, 166
Hirado, 35
History, 1-2, 4, 13, 18-19, 24, 26, 29, 33-36, 41, 47, 49-50, 68, 70-71, 80, 82-83, 106-107, 123, 126, 128-129, 148, 150, 159, 163-164, 167, 169, 172-173, 176, 178, 184-187, 196, 200, 202, 205
Hodges, William, vii, 58-59, 64-66, 69
Holocaust, iv-v, 1, 3-4, 13-14, 16, 19, 22, 24, 26-27, 202
Hosseini, Khaled, v, 5, 70-71, 74, 76-79, 83-84
Hugo, Victor, 142
Hulme, Peter, 2, 9, 202, 205
Hunting, 7, 109, 111, 113-114, 121, 129, 157, 161, 188, 190-195, 198, 200
Huysmans, Joris-Karl, 138, 144

I

Imperialism, 2, 5-6, 42, 47, 78, 83, 87, 106, 189-190, 193, 199-200
Impressionism, 6, 142, 144-146, 148
India, 2, 6-7, 9, 41-42, 64, 66-67, 69, 151-152, 156, 163, 166, 190, 192, 194-200, 203-205
Islam, 78, 84, 150, 154, 164
Israel, 25, 73
Italy, 27, 154, 166

J

Japan, v, 4, 35-40, 42, 44-50, 191
Johnson, Lionel, 112, 114, 127-128, 146
Joppien, Rüdiger, 58, 69

Index

K

Kapor, Momo, 96-101, 103-104, 107
Keysler, Johann Georg, 61, 69
Kipling, Rudyard, 81, 83, 192, 194
Kojève, Alexandre, 134
Kostantinov, Aleko, 85
Kušan, Ivan, 97-99, 102-103, 105

L

Le Bruyn, Cornelius, vii, 53-57, 60-61, 68
Le Comte, Louis Daniel, 61, 68
Lessing, Gotthold, 61, 63, 69
Levant, 151-152, 154-155, 157-158, 163, 165-167
Lévi-Strauss, Claude, 1
Literature, 2, 5, 7, 9, 26, 49, 68, 74, 81-82, 97, 106, 121, 127, 136, 139-140, 142, 148-149, 168, 175, 189, 194, 199, 205
London, vi-viii, 6-9, 30-33, 49-50, 54, 57, 59, 62, 68-69, 79, 82-84, 106-114, 116, 118-121, 123-124, 126-129, 133, 135, 137-149, 155, 164, 166-167, 175, 177, 186-187, 191, 200, 203
London, Jack, 129

M

MacKenzie, John M., 119, 193, 195, 198, 200
Magazine cartoons, 188
Mahomet, Dean, 66-67, 69, 167
Malaita, viii, 116-121, 127
Mallarmé, Stéphane, 138, 144
Manet, Édouard, 140
Martineau, Harriet, 188-189
Masculinity, 5-6, 85-87, 89-90, 92-93, 95-97, 99, 103-105, 193
Maundrell, Henry, 152-154, 158, 165, 167
McCrum, Mark, 168-173, 176-178, 181-184, 186-187
Medhurst, W. H., 42
Meiji Restoration, 37, 39, 50
Melanesia, vi, 1, 6, 108-109, 113-114, 119, 122-124, 126, 128-129
Memorial to the Murdered Jews of Europe, 13, 24, 27
Memorials, 3, 8, 14-16, 24, 26-27, 29, 179, 187, 202
Memory, iv-v, 1, 3-5, 11, 13-15, 17-19, 21-27, 30-31, 33-34, 49, 146, 202, 205
Memory travels, 202
Mendès, Catulle, 142
Middle East, 3, 5, 70, 73-76, 78-84, 150, 158, 167
Mills, Sara, 87, 107, 189, 200
Missionary, viii, 35, 38-39, 41, 119, 123, 152, 160, 167
Moabit, 15-16, 25
Mobility, 2, 74, 76, 201-202, 204-205
Mobility regimes, 202, 204-205
Moore, George, 6, 133, 136, 138-143, 148-149
    Confessions of a Young Man, 133, 149
    Modern Lover, A, 141
Morrison, Toni, 4, 28-30, 34
Muirhead, William, 38, 41
Museum, 3, 13-14, 29, 34, 175, 203
Muslims, 78, 150, 153-154, 156, 159-160, 163-166

N

Nagasaki, 35-37, 44
Nakamuda Kuranosuke, 38, 40, 45-46, 49-50
Nancy, Jean-Luc, 134-135
Nanjing, 35, 38
Nazi regime, 16
New Woman, 189-190
New Zealand, vii, 59, 65
North America, 36

O

Opium War, 35, 43, 45, 49
ordinariness, 168, 171-173, 180, 183-184, 187
Orientalism, 2, 7, 9, 76, 82-84, 86, 96, 104-107
Osaka, 37, 49
Otherness, v, 1, 4, 6-7, 51, 53, 60, 66-67, 87
Ottoman Empire, 151, 154, 165, 167

P

Pakistan, 73, 79
Palestine, 3, 9, 72, 83

Paris, vi, 1, 6, 8, 24, 34, 98, 100-101, 133, 138-140, 142-144, 148-149
Parsons, Abraham, 155-157, 167
Pater, Walter, 139, 149
Paton, Andrew Archibald, 162, 166-167
Perry, Matthew, 35
Picart, Bernard, vii, 53-55
Piranesi, Giambattista, 60
Postcolonial, 76-77, 81-84, 107, 168, 184-185, 201
postcolonial travel writing, 185
Potolsky, Matthew, 134, 149
Pound, Ezra, 148
Pratt, Mary Louise, 2, 4, 7, 9
Protestantism, 152-153, 165
Pulitz Bridge Deportation Memorial, 4-8, 16

Q

Qing Dynasty, 38, 41, 45
Queen Elizabeth I, 151
Queen Victoria, 188
Queensland, 118-119, 128, 180

R

Rayner Parkes, Bessie, 188
Redon, Odilon, 142
Representation, 1-6, 9, 55, 57-58, 60, 66, 77, 84, 107, 137-138, 140, 143, 146-148, 169, 185
Ridley, Glynis, 203
Robinson, Jane, 167, 191, 194, 200
Rome, 27
Russell, Alexander, 9, 155, 159-161, 165-167
Russia, 55, 74, 97, 151, 191

S

Sachs, Nelly, 19, 26
Said, Edward, 9, 84
Saudi Arabia, 73
Savory, Isabel, 189, 192-200
Scotland, 134
Senzaimaru, 4, 36-37, 39-41, 44-45, 47-50
Serbia, 97
sexual revolution, 189
Shanghai, 35-50
Shelley, Percy Bysshe, 144

Shoah, 3, 14, 18
Sociology, 1, 3, 106
Sontag, Susan, 64, 69, 116, 129
South America, 191, 196, 198
Sparrman, Anders, 56, 69
Spivak, Gayatri Chakravorty, 91, 98, 107
Sport, 164, 188, 190, 192, 194, 198, 200
Stewart, Rory, 72-74, 81, 84
*Stolpersteine (*Stumbling Stone*),* vii, 19-22, 24, 27
Swift, Jonathan, 6, 56, 69, 136, 142
Symons, Arthur, 14, 133, 136, 142-149
  Colour Studies in Paris, 142, 149
  London Nights, 145-146, 149
  Silhouettes, 17, 145-146, 149
Syria, 7, 150-152, 155-156, 159, 165-167

T

Takasugi Shinsaku, 37, 40, 45, 48-50
Terra del Fuego, 57
Tokugawa Dynasty, 35, 41, 55,
Tourist novel, 76-77, 82
Trail, v, 1, 3, 13-14, 16, 18-19, 22, 24-25, 30, 33, 180
Treaty of Kanagawa, 35
Turkey, 73, 79, 150, 156, 167
Turner, Edith, 9
Turner, Victor, 1, 174, 187

U

United Kingdom, 7, 31, 33
United States of America, 169, 173-174

V

Van Egmond, Jan Aegidius, 153
Verlaine, Paul, 138-139, 142, 144
Vernon, Francis, 161-163, 167
Virus, 205

W

Walls. 13, 18, 156
Warsaw Pact, 97
Wilde, Oscar, 137
Winterbottom, Michael, 71, 74, 79, 84
Wolfreys, Julian, 138, 149
women's health, 194, 198
Wordsworth, William, 141

Index

X
Xiucheng, Li, 38, 42, 44
Xiuquan, Hong, 38

Y
Youngs, Tim, 2, 9
Yugoslavia, 87, 96-97, 99, 102, 104, 106-107

Z
Zola, Émile, 139-140

www.ingramcontent.com/pod-product-compliance
Lightning Source LLC
Chambersburg PA
CBHW072153100526
44589CB00015B/2211